GEOGRAPHERS
Biobibliographical
Studies

VOLUME 16

GEOGRAPHERS: BIOBIBLIOGRAPHICAL STUDIES

GEOGRAPHERS
Biobibliographical Studies

VOLUME 16

Edited by
Geoffrey J. Martin

on behalf of the
Working Group on the History of Geographical Thought
of the International Geographical Union and the
International Union of the History and Philosophy of Science

MANSELL

First published 1995 by
Mansell Publishing Limited, *A Cassell Imprint*
Villiers House, 41/47 Strand, London WC2N 5JE, England
387 Park Avenue South, New York, New York 10016-8810, USA

British Library Cataloguing-in-Publication Data
Geographers: biobibliographical studies – Vol. 16
   1. Geographers – Biography – Periodicals
   910'92'2   G67

   ISBN 0-7201-2209-0

Typeset by Litho Link Ltd, Welshpool, Powys, Wales
Printed and bound in Great Britain by Biddles Ltd, Guildford and King's Lynn

Contents

Introduction

This volume of *Geographers* contains eleven biobibliographical studies. Three derive from France, two from England, and one each from Austria, China, Korea, Norway, Peru (by adoption) and Poland. One of the figures is taken from the sixteenth century, two from the nineteenth century, five from the *fin-de-siècle*, and three from the twentieth century.

And so we have a *mélange* – of pre- and post-disciplinary figures, of western and non-western persons, and of university and non-university careers. When they are elaborated all too briefly in *Geographers*, as subjects of biobibliography, certain common themes and strands may become apparent in their life paths. In part that is the product of, and encouraged by, the three fundamental headings which we have adopted since the inauguration of the series: Education, Life and Work; Scientific Ideas and Geographical Thought; and Influence and Spread of Ideas (these three main headings couple with end matter which includes bibliography, sources and chronology). Inevitably more is written about life than about thought and spread of ideas. All of this encourages us to learn what is common to the geographical enterprise and what, perhaps, is less common, though nevertheless helpful to the structure of contribution.

The biobibliographies of our series constitute an inexhaustible supply of building blocks in the history of geographical thought – inexhaustible because eminent geographers are passing away in aggregate more quickly than they can be elaborated on and appreciated in *Geographers*. Each of these essays functions as a corrective to the belief that geographers share a group-like mentality and move in a pack. The individuals who have helped make geography what it is are many, and they have wrought their contribution on a large canvas. From the accumulated anthology of writings and the lexicon of achievement one recognizes (or at times perceives) a number of national geographies, and a larger, much more loosely knit and hard-to-define international geography.

The internationalization of our field may be traced through the work of a number of travellers, intellectuals and societies. Arguably, however, there was a more formal beginning to this process when in 1871 in Antwerp, members of our guild assembled to commemorate and honour Mercator and Ortelius. Meetings followed in a series of international geographical congresses: in Paris (1875), Venice (1881), Paris (1889), Bern (1891), London (1895), Berlin (1899), Washington, DC (1904), Geneva (1908) and Rome (1913). Hostilities interrupted those quadrennially attempted meetings and in 1922 the International Geographical Union was founded. This latter gave birth to commissions, working groups, and meetings both regional and international. And just as institutions have contributed to the development of an international science, so we can recognize individuals who have contributed similarly.

It is perhaps invidious to select an individual representative of the internationalization of geographical science. Nevertheless, and in remembrance of an honorary member of this commission, we might select as exemplar Richard Hartshorne (1899–1992), as one who attempted to trace the development and explicit nature of (western) geography. His work, *The Nature of Geography* (1939), was translated into Spanish, German (partial), Chinese and Japanese, though other workers in non-English-speaking lands sought to read parts of the study in the original publication. It was a large contribution which aimed to find common ground in the nature of our undertaking, and was one of the first books concerning this task to experience global diffusion. That work might yet be subject to a bibliobiography. To a lesser extent all subjects of the published biobibliographies contribute to the understanding of a shared science. Their work renders them internationalists as copies of *Geographers* are distributed, shared and read around the world.

This volume, as mentioned above, contains eleven biobibliographical entries. A few lines by way of introduction: Xu Hongzu has provided a description of much of early seventeenth-century China that remains of interest to contemporary scholars. He also demonstrated and brought to a new level the method of systematic observation of natural phenomena and resultant geographical condition. Hongzu's remarkable work was, at least in part, brought to the west by the Italian P. Martini. Also from east Asia (Korea), Kim demonstrated the reality and the worth of bringing the regional geography text and map of that area into juxtaposition as an inseparable pair. The third of what might be called our pre-disciplinary subjects is the Italian Antonio Raimondi del Acqua, who adopted Peru as his own land and proceeded to write of his travels and map the country from his own gathered data. He has bequeathed to us the four-volume *El Perú*, a virtual geographic handbook of the country.

Five of the subjects in this volume lived at least some of their professional life at or near the turn of the nineteenth century – a period in the western world when learning, institutions and professionalism were melding to form discipline. Paul Reclus, nephew of Elisée, curiously stood apart from mainstream activity and remained aloof from helping with the emerging discipline. He was extremely modest, was given to unconventional enthusiasms, and devoted much of his time to compiling the works (and promoting the reputation) of his uncle, Elisée Reclus. Louis Cuisinier was grandson of Elisée Reclus, an archivist by sympathy, earth scientist by capacity, and geographer by family affiliation. He became very knowledgeable on Indochina and especially what might be called mining geography. He wrote considerably on mining potential in the tropics, and developed maps of these areas, little known because they have been retained as the property of these companies. The third Frenchman, Paul Meuriot, was not related to the Reclus family. He was a keen and knowledgeable student of social statistics, who exploited his data to study urbanization in all of its modes. In some ways his work was much ahead of its time; in part it was overshadowed by the Golden Age of French geography when the study of the region, dominated by the students of Vidal de la Blache, was at its zenith. Never forgotten, his work was variously discovered and rediscovered. J.L. Myres of England, a classicist associated with Oxford University, developed an impressive knowledge of ancient geography, and published some excellent work in geographical history, typified by *The Dawn of History* (1911). In insisting on the primacy of the mind, and examining the interaction of the mind with the external environment, author Koelsch suggests that Myres anticipated cognitive humanistic geography. And the Norwegian Nansen, crypto-geographer, developed a first-hand knowledge of the Arctic and oceanography. His work with the *Fram* ('a floating scientific station') and his discoveries with regard to the interior of Greenland contributed to understanding of the Polar Basin and climatic theory. It was as a result of the three-year drift of the *Fram* that the 'Ekam Spiral' was developed. His Greenland crossing, embracing a winter spent with the Eskimo at Godthåb, resulted in *Eskimo Life*. Other than that, Nansen's remarkably productive life brought attention to the northlands (Norway in particular) and the harsh reality of the northern physical environment.

The three remaining geographers written of in this volume are from the twentieth century: Franciszek Bujak, John N.L. Baker and Hans Bobek. Bujak, of Poland, devoted his life to a study of history and geography with special reference to Poland. He also became very interested in the history of cartography and geographical science, and, indeed, in the whole of Polish intellectual history. His dissertation, entitled 'The Development of Geography in Poland in the Middle Ages and at the Beginning of the Sixteenth Century', has been the subject of reference frequently and to the present time. Bujak, however, was prolific and published some four hundred items. J.N.L. Baker, who spent most of his life at Oxford University, also read history and geography. As was the case with Bujak, he too had a keen interest in historical geography and the history of geography. In 1963 his geographer friends and colleagues at Oxford gathered together a selection of his writings which were published as *The History of Geography: Papers by J.N.L. Baker*. This, like the rest of his work, was widely read throughout the English-speaking world. Finally the Austrian Hans Bobek completes the roster for Volume 16. Author Lichtenberger informs us of this talented geographer who grew up and spent his life in the German-speaking lands. Bobek, emphatic believer in the concept of '*landschaft*' and pioneer in social geography, geo-scientist and positivist, lived to witness the radical departure brought about at the Kiel meeting of German geographers in 1969. He was not enamoured of the new and numerate approach, but at the age of 65 began the study of statistics to keep apace of the new movement. His knowledge of Iran, derived from fieldwork, and his theory of rent capitalism inspired in the field are particularly of note.

The editor of this publication continues to invite further biobibliographies of eminent and deceased geographers.

Geoffrey J. Martin
Editor

List of Abbreviations

Forsch. z. Dt. Landes- u. Volkskunde
Forschungen zur Deutschen
Landes- und Volkskunde

Geog. Teacher Geographical
Teacher

Geogr. J. Geographical Journal

Jb. d. Geol. Bundesanstalt Wien
Jahrbuch der Geologischen
Bundesanstalt Wien

J. Journal de la Société de
Statistique de Paris

Kwart. Hist. Kwartalnik
Historyczny

Mémoire Acad. des sc. Mémoire de
l'Académie des Sciences morales et
politiques

Mitt. d. Österr. Geogr. Ges.
Mitteilungen der Österreichischen
Geographischen Gesellschaft

Mitt. Geogr. Ges. Wien Mitteilungen
der Geographischen Gesellschaft
Wien

Prace geogr. Prace Geograficke

Proc. Class. Assoc. Proceedings of
the Classical Association

*Rozpr. Akadem. Umiej. Wydz. hist.-
filozof.* Rozprawy Wydziału
Historyczno-Filozoficznego
Akademii Umiejętności

Scot. Geogr. Mag. Scottish
Geographical Magazine

South African Geogr. J. South
African Geographical Journal

Wiadom. Numizm.-Archeol.
Wiadomosci Numizmatyczno-
Archeologiczne

John Norman Leonard Baker

1893–1971

Robert W. Steel

1. Education, Life and Work

John Norman Leonard Baker – J.N.L.B. to many of his friends – was born at 33 Beaumont Street, St Clements, Liverpool, on 12 December 1893 and died in Oxford after a short illness on 16 December 1971. He was the first son of a clergyman of the Church of England, the Rev. John William Baker, whose livings were mainly in Lancashire. Baker grew up in the vicarage in Shaw Street, Liverpool, very near to the Liverpool Collegiate School. The parish adjoined the area now extensively occupied by the University of Liverpool, whose Victoria Building on Brownlow Hill was opened in 1892, the year before Baker's birth. Baker approved heartily of the words on a plaque on that building that states that 'For the advancement of learning and ennoblement of life the Victoria Building was raised by men of Liverpool in the year of Our Lord 1892.'

Baker's school education was divided between Liverpool Collegiate and Liverpool College, to which he moved in 1911 and on whose governing body he subsequently served for many years. He left the college in 1913, having been elected in January to an Open Exhibition in Modern History (worth £50 per annum) at Jesus College, Oxford; thereafter, apart from war service (in two world wars) and a period of a few months when he taught in the University of London, his life was based on the college as an undergraduate, a postgraduate student, a lecturer, a fellow and bursar, and finally an emeritus fellow. Devotion to Jesus College and service to the City of Oxford, of which he was a city councillor from 1945 to 1967 and Lord Mayor in 1964–65, were among Baker's most marked qualities during much of his adult life.

As an undergraduate he read the Honour School of Modern History, which gave him the discipline and the historical sense that were so important in his career as a geographer. His studies were interrupted, though also enriched, by the First World War, for he joined the army in December 1914. He was a second lieutenant in the King's Liverpool Regiment and served in France, where he was severely wounded on the Somme and where he owed his life to the courage and concern of a French infantryman. During a long period of convalescence he married Phyllis Marguerite Hancock on 8 November 1917; as a direct consequence of his marriage he had to vacate the Open Exhibition that he had been awarded in 1913. Two years in the Indian Army followed, and as a lieutenant in the 119th Indian Infantry he developed that great interest in the land and its people that never left him and about which he published some important papers. Returning to Oxford, he completed his studies, reaching distinction standard in the Modern History School in 1920. He then embarked on the one-year course for the Diploma of Geography, which he obtained in 1921 with distinction, and was awarded the Herbertson Memorial Prize. While an Assistant in Geography during the academic year 1921–22 he undertook the research for which in 1922 he was awarded the degree of B.Litt. In his thesis 'Geographical Aspects of the Peninsular War', he combined historical perspective with his newer geographical insights.

In 1922–23 he was Lecturer in Geography in the small Department of Geography at Bedford College, University of London. From there he returned in 1923 to Oxford on appointment as Assistant to the Reader in Geography (the Reader was H.O. Beckit). The Oxford School of Geography had very few staff members but was growing significantly throughout the 1920s, largely because of the increasing numbers of graduates who, as Baker himself had been, were interested in taking the diploma course in geography, particularly if they planned careers in schoolteaching. There were also important (and time-consuming) summer schools, patronized almost entirely by teachers, many of whom subsequently played an important part in the building up of geography as a school subject during the inter-war years. Baker inevitably was caught up in heavy teaching duties as well as administrative and other responsibilities in the School of Geography, in which he was appointed University Lecturer in Geography for five years from 1 October 1927. These preoccupations never prevented the research worker that he was from publishing scholarly papers, particularly those that traced the history of geography, especially in Oxford, and the progress of geographical discovery. In 1931 he published his *History of Geographical Discovery and Exploration*, a standard work for many years and one that is still consulted more than six decades after its first appearance.

Baker played a very active, though often behind-the-scenes, role in the campaign that led ultimately, after surmounting many obstacles and encountering much opposition, to the establishment of the Honour School of Geography and the simultaneous creation of the Professorship of Geography in 1933. During 1931 both his senior colleagues in the school died – H.O. Beckit in February at the age of 56 and J. Cossar in September aged 49, and in consequence Baker carried many additional burdens in both administration and teaching. In the academic year 1931–32 he was appointed Acting Director of the School of Geography. Had he then been appointed to the newly established chair the course of his own life and the progress and direction of the subject in Oxford might have been very different.

The Electors appointed Kenneth Mason of the Survey of India to the Chair in 1933, and Baker became Reader in Geography (Reader in Historical Geography from 1935). He was also appointed Lecturer in Geography in his College, Jesus, which shortly afterwards offered a Welsh Scholarship or an Open Exhibition to would-be geographers (though inevitably this was in competition with Modern Languages). A few years later, in 1939, he was elected to a Fellowship at Jesus with the title of 'Tutor and Senior Bursar', and he continued in that office, as Official Fellow and Bursar, until his retirement in 1962, his fellowship having been extended for a year. For most of the period he was the sole bursar, despite his original title, with responsibilities for both the Home and Estates side of the College's life. His very special contribution to the progress of Jesus College could never have been made had he been appointed to the Chair in 1933, although his direct and positive involvement in the development of geography in Oxford continued for many years, even after his retirement from his University Readership in 1947.

During the Second World War Baker volunteered for service with the Royal Air Force and was engaged in intelligence duties in London, Oxford and elsewhere, before demobilization and return to the bursarship of his college in Oxford.

Baker was a great supporter of the work of a number of learned societies. He was for many years a Fellow of the Royal Geographical Society, of whose Council he was a member from 1946 to 1948. But his special contribution to the progress of the subject was through the Institute of British Geographers, established in 1933, of which he was a founder member. He served on its Council during the Institute's early formative years, became its secretary in 1936, and was one of the officers (with R.O. Buchanan) who watched over the infant society during the Second World War, having been elected President in 1939. He remained in office until the Institute resumed its activities after the end of the war, when he presided over the first of the post-war meetings in 1946. He was intimately involved, as the Bursar of Jesus College, in two of the first residential conferences arranged by the Institute in Oxford in 1949 and 1951.

He was a regular attender at the annual meetings of the British Association for the Advancement of Science, and he strongly supported the arrangement whereby members of Section E (Geography) combined with members of the Institute of British Geographers in field excursions (following the Association's annual meeting) each summer. He became Recorder of Section E in 1936 and remained in office throughout the war years until 1949. In 1955 he was President of Section E at the meeting held in Bristol.

Baker's interests as a geographer and as an historian found expression in his involvement in the Hakluyt Society, of which he was a lifelong member, from 1924 until 1971. He served the society in numerous ways, being both trustee and treasurer from 1946 until the time of his death, and president between 1955 and 1960.

Baker's service to the community outside the University of Oxford included work for bodies such as the Oxford Delegacy of Local Examinations and the Oxford Preservation Trust, but it was dominated by his work as a councillor of the City of Oxford. He was first elected – by the bursars of the colleges – as a city councillor in 1945. He became an alderman in 1963. Oxford had been raised in status to a city with a lord mayor in 1962, and Baker was the first university representative to be honoured as lord mayor, an office that he held in 1964–65. He retired from the council in 1967.

2. Scientific Ideas and Geographical Thought

Baker was no theoretician and had little sympathy with many developments in the subject in post-war years, particularly those that dominated thinking in geography from 1960 onwards. He was especially suspicious of many aspects of the quantitative revolution and of those whose main concern seemed to be 'number crunching'. Indeed in his latter years the historian in him became increasingly marked, and his last publication, produced only a few months before his death, was the history of his college, *Jesus College, Oxford, 1571–1971*. This book of 153 pages, published at the time of the college's quatercentenary, was in fact only a small version of the work that he had done but which, in his own words, 'proved to be too long and could not be printed in time for the Celebrations of 1971', so that 'very drastic revision was necessary'.

His initial training as an historian also showed itself in his devotion over many years to the work of the Hakluyt Society. Significantly, the citation for the prestigious Victoria Medal, awarded to him in 1964 by the Royal Geographical Society, referred to his 'contributions to the history of geography'.

He was widely read in both history and geography and encouraged his pupils to study not only standard texts but also related work, in French and German as well as in English. He was one of the few geographers between the two world wars who appreciated the significance of Sir Halford Mackinder's *Democratic Ideals and Reality: A Study in the Politics of Reconstruction*. He had commended this to his pupils long before the book was reissued (in 1942) to be available to students of geopolitics in the USA and in Britain, who were thus introduced to Mackinder's thesis:

> Who rules East Europe commands the Heartland;
> Who rules the Heartland commands the World-Island;
> Who rules the World-Island commands the World.

Baker's presidential address to the Institute of British Geographers in 1947, entitled 'Geography and Politics: The Geographical Doctrine of Balance', summarises many of his ideas. It is introduced by a long quotation from Shakespeare's *King Henry IV*, Part 2, and concludes with some wise words from Edmund Burke; and this is followed by Baker's lengthy footnote referring to British geography's 'severe loss [since his address had been delivered] in the death of one of its greatest exponents, Halford John Mackinder'. Baker's final comment is his hope 'that what I have written would have met with the approval of a great master, whose inspiring words and stimulating ideas have done so much to advance the cause of British geography'. Many of Baker's pupils would feel that a similar tribute should be paid to Baker himself.

3. Influence and Spread of Ideas

Baker's influence on his many students in Oxford was tremendous albeit intangible. While he never suffered fools gladly and could identify humbug or half truths with remarkable ease, he was for most of his pupils outstanding as a tutor. To this day he is remembered by them with respect and affection. His role was never confined to the hearing of essays or the stimulation of critical thought. He never liked the word 'teaching' applied to undergraduates. 'You cannot (or should not) teach a man,' he often said, 'only encourage him to teach himself.' He was never head of a university department of geography but for many years, and particularly in the early years of the Honour School of Geography in Oxford during the 1930s, 'Baker' and 'Oxford geography' were almost synonymous terms. During that period he taught a very high proportion of the students in the School of Geography – including all those who were awarded first class honours degrees – and his influence on them and on others whom he taught subsequently, especially in his own college, Jesus, was considerable. His organization each year of a foreign field excursion, usually in France or in Germany, established a very lively tradition of fieldwork overseas in the Oxford School and ensured that he came to know well a fair proportion of the undergraduates in each year in the School. Appreciation of the range of his interests and the extent of his influence on his pupils were indicated in the Festschrift-type volume, *The History of Geography: Papers by J.N.L. Baker*, published in 1963 to mark his retirement in Oxford. It was masterminded by A.F. Martin with the assistance of other Oxford colleagues, all Baker 'disciples': Frank Emery, Rowley Gullick, Mary Marshall, Ernest Paget and Robert Steel. It was, they wrote, 'some public and, so to speak, ceremonial expression both of their regard for their old tutor and of their admiration for his services to British geography, and particularly to geography in Oxford'.

Baker's influence on the progress of his discipline expressed itself particularly through his close involvement with the Institute of British Geographers, of which, as already noted, he was a founder member and subsequently secretary and then president. He was in fact one of those whose faith in its future made its post-war revival a certainty. Throughout his life he maintained close links with the Royal Geographical Society, with its special interest in exploration and discovery. Although he never held national office in the Geographical Association, he regularly attended its London conferences, particularly when they were held simultaneously with the early annual meetings of the Institute of British Geographers. He was a stalwart member of the Oxford Branch of the Association for many years, supporting it whenever possible by attendance at its meetings. He also gave regular and much appreciated support by his presence at many of the meetings of the Oxford undergraduates' society, the Herbertson Society.

Through his work for the Hakluyt Society he had an impact on a still wider circle of scholars, and his reputation was always very high with the historians as well as the geographers who were among its members.

Those who knew him well turned repeatedly to Baker for counsel and advice right to the end of his life, though his direct influence on the progress of geography in Britain waned inevitably as a consequence of his increasing involvement in university, college and civic affairs in Oxford. Yet his interest in his subject – and his former pupils – remained unabated, and his study of the literature continued throughout his life, although he was cautious about many of the new developments in the subject during the 1960s, and highly critical of much that was being said and done in the name of geography at that time. In both the University and the City of Oxford Baker was highly esteemed for his bursarial skills, his shrewd judgement, and his down-to-earth and forthright views on a wide range of subjects.

Baker was rightly proud of the acheivements of many of his pupils, especially those from his own college, several of whom became professors of geography and heads of departments. He showed equal pleasure and pride in his two daughters, Janet and Rosalind. Both were evacuated from England to the USA during the Second World War and on their return became graduates of the University of Oxford. Janet read the PPE Honour School (Philosophy, Politics and Economics) at St Anne's College and Rosalind the Honour School of Geography at Lady Margaret Hall. Janet followed her father's interest in political life, first as a councillor and later as an alderman of the City of Oxford, and then nationally after becoming a life peer in 1971. Among her appointments she was Minister of State in the Department of Education and Science, Deputy Foreign Minister, and Leader of the House of Lords. Rosalind, like her father, became a geography teacher and taught in schools in London and, with M.E. Johnstone, published a well-known school textbook on the British Isles.

Bibliography and Sources

1. OBITUARIES AND REFERENCES ON JOHN NORMAN LEONARD BAKER

Anon., 'J.N.L. Baker', *Jesus College Record* (1971), 13–14. At p. 22 there is a photograph (black and white) of a portrait of Baker by M.P. Holland, presented to him by the Bursars of the Oxford colleges.

Cross, Rev. L.B., address at memorial service for J.N.L. Baker, 29 January 1972, *Jesus College Record* (1971), 20–1.

Steel, R.W., obituary, *Geographical Journal* Vol. 138 (1972), 276–7.

Steel, R.W., 'J.N.L. Baker: a personal view', *Jesus College Record* (1992/1993), 22–7.

2. BIBLIOGRAPHY OF WORKS BY JOHN NORMAN LEONARD BAKER

The History of Geography: Papers by J.N.L. Baker Presented to Him by His Pupils was published in 1963 by Basil Blackwell in Oxford. Six of his former pupils (C.F.W.R. Gullick, A.F. Martin, R.W. Steel, Mary Marshall, E. Paget and F.V. Emery) were responsible for the selection and editing of fifteen of his published papers, to which three hitherto unpublished contributions were added. The latter are listed with the date 1963 in the following bibliography. The fifteen other papers included in the collection are indicated by an asterisk. Reviews have been omitted but review articles (of which there are six) are included.

a. The History of Geography and Exploration and Historical Geography

   1926   'Some sources for the history of exploration', *Geographical Teacher*, Vol. 13, 307–16.

* 1928   'Nathanael Carpenter and English geography in the seventeenth century', *Geographical Journal*, Vol. 71, 261–71.

* 1931   'The geography of Daniel Defoe', *Scottish Geographical Magazine*, Vol. 47, 257–69.

   1931   *A History of Geographical Discovery and Exploration*, London, 544 pp. New edition 1937. Translated into French and published in Paris, 1949. Translated into Polish and published in Warsaw, 1959.

1933   Contributions to *Catalogus Mapparum Geographicarum ad Historiam Pertinentium*, Warsaw.

* 1935   'Academic geography in the seventeenth and eighteenth centuries', *Scottish Geographical Magazine*, Vol. 51, 129–43.

1936   'England in the seventeenth century', in H.C. Darby (ed.), *An Historical Geography of England before A.D. 1800*, Cambridge, 387–443.

* 1936   'The last hundred years of historical geography', *History*, NS, Vol. 21, 193–207.

1936   'Some original maps of the East India Company', *Tijdschrift van het Koninklijk Nederlandisch Aardrijksundig Genootschap*, second series, Vol. 53, 665–7.

* 1937   'Some Dutch charts of the seventeenth century', *Imago Mundi*, Vol. 2, 16.

* 1937   'The earliest maps of H. Moll', *Imago Mundi*, Vol. 2, 36.

1938   'Medieval trade routes', *Historical Association Pamphlet* 3. Reprinted, with corrections, in G. Barraclough (ed.), *Social Life in Early England*, London, 1960.

1938   Four chapters in Sir P. Sykes (ed.), *The Story of Exploration and Adventure*, London.

1940   'The problems of Vinland', *Geographical Journal*, Vol. 96, 48–50 (review article).

* 1944   'Sir Richard Burton and the Nile sources', *English Historical Review*, Vol. 59, 48–61.

1946   Four chapters in Lucien Mazenod (ed.), *Les Explorateurs Célèbres*, Geneva.

1947   'James Cook (senior scout patrol names)', *The Scouter*, May 1947. Reprinted in *Patrol Books*, No. 3, Part 2, 1949.

* 1948   'Mary Somerville and geography in England', *Geographical Journal*, Vol. 111, 207–22.

1950   Contributions to *Chambers' Encyclopaedia*, including 'Exploration after 1415', 'Livingstone' (in part), 'Speke', 'Foa', 'Frobisher', 'Gilbert', 'Grenfell', 'Varenius', 'Wissmann' and others.

* 1952   'The development of historical geography in Britain during the last hundred years', *Advancement of Science*, Vol. 8, 406–12.

1954   (with Audrey M. Lambert) 'The seventeenth, eighteenth and nineteenth centuries', in A.F. Martin and R.W. Steel (eds.), *The Oxford Region: A Scientific and Historical Survey*, Oxford.

* 1955   'Geography and its history', *Advancement of Science*, Vol. 12, 188–98.

* 1955   'The geography of Bernhard Varenius', *Transactions and Papers, Institute of British Geographers*, Vol. 21, 51–60.

1957 'The journal of Captain James Cook, Vol. 1', *English Historical Review*, Vol. 72, 133–6 (review article).

1960 'The ends of the earth', *Geographical Journal*, Vol. 126, 70–3 (review article).

1961 'The exploration of Central Africa', *Geographical Journal*, Vol. 127, 228–9 (review article).

1962 'The exploration' in *The Nile Quest. Centenary Essays and Catalogue*, Kampala, Uganda Museum.

1962 'John Hanning Speke', *Geographical Journal*, Vol. 128, 385–8.

1963 'Major James Rennell, 1742–1830, and his place in the history of geography', in *The History of Geography*, 130–57.

1963 'The history of geography in Oxford', in *The History of Geography*, 119–29.

1963 'Mythical lands in history', in *The History of Geography*, 179–87.

b. India

1923 'The economic future of India', *Geographical Teacher*, Vol. 12, 127–31.

1928 'Notes on the natural regions of India', *Geography*, Vol. 14, 447–55. Reprinted in *Spence Training College Annual*, Jabalpur, 1930.

* 1928 'Some geographical factors in the campaigns of Assaye and Argaon', *Army Quarterly*, Vol. 17, 44–53.

1934 'The 1931 census of India', *Geographical Journal*, Vol. 84, 434–7 (review article).

1936 'Some problems of population in India', *Scottish Geographical Magazine*, Vol. 52, 231–40.

1954 'The regions of India and Pakistan', *Geographical Journal*, Vol. 120, 493–6 (review article).

c. Jesus College, Oxford: Its History and Its Buildings

1944 'Captain Angus Buchanan, V.C., M.C.', *Jesus College Magazine*, Vol. 6, No. 76, 184.

1947 'The chapel', *Jesus College Magazine*, Vol. 7, No. 86, 13–14.

1948 'Vale' (on J.G. Edwards), 'The college buildings', *Jesus College Magazine*, Vol. 7, No. 88, 54–5.

1948 'Wars in college history', *Jesus College Magazine*, Vol. 7, No. 89, 70–2.

1949 'Lambeth 1685–1949', *Jesus College Magazine*, Vol. 7, No. 90, 94–6.

1949 'The college in the long vacation', *Jesus College Magazine*, Vol. 7, No. 92, 14–15.

1950 'Jesus College and Boston University', *Jesus College Magazine*, Vol. 7, No. 93, 2–3.

1950 'The chapel windows', *Jesus College Magazine*, Vol. 7, No. 94, 12–13.

1951 'From the College Minute Book – Hilary Term 1831 and 1862', *Jesus College Magazine*, Vol. 7, No. 96, 7–8.

1951 'November and May. From the College Minute Book', *Jesus College Magazine*, Vol. 7, No. 97, 9–10.

1951 'Fifty years ago. From the College records', *Jesus College Magazine*, Vol. 7, No. 98, 19–21.

1952 'The Bursar looks back', *Dragon. Jesus College Magazine*, Vol. 7, No. 100, 5–6.

1953 'The Long Vacation 1953', *Dragon. Jesus College Magazine*, Vol. 7, No. 104, 4–5.

1954 'A memory of J.R. Green', *Dragon. Jesus College Magazine*, Vol. 7, No. 105, 4.

1954 'An undergraduate of 1774–5', *Dragon. Jesus College Magazine*, Vol. 7, No. 107, 3–4.

1956 '1856 and 1906', *Dragon. Jesus College Magazine*, Vol. 7, No. 113, 8–9.

1958 'Obituary' (obituary notices of Canon Maurice Jones and of D.L. Chapman), *Dragon. Jesus College Magazine*, Vol. 8, No. 1, 23–4.

1959 'From the Bursary. Hugh Jones – father and son', *Dragon. Jesus College Magazine*, Vol. 9, No. 1, 4–8.

1960 'From the Bursary. James Matthews and his collected papers', *Dragon. Jesus College Magazine*, 43–4.

1962 'In the Long Vacation', *Dragon. Jesus College Magazine*, 15–19.

1962 'The College buildings', *Jesus College Record*, 14–16.

1963 'The Missionary Fellows', *Jesus College Record*, 24–30.

1964 'The College site', *Jesus College Record*, 15–20.

1964 'Thomas Pardo', *Jesus College Record*, 20–4.

1965 'Andrew Hughes Matthews', *Jesus College Record*, 17–25.

1966 'Edmund Meyricke and his benefaction', *Jesus College Record*, 19–28.

1967 'Sir Leoline Jenkins', *Jesus College Record*, 18–23.

1968 'The Principal's Lodgings', *Jesus College Record*, 20–1.

1968 'Francis Mansell and his inventory', *Jesus College Record*, 21–8.

1969 'The poor and needy', *Jesus College Record*, 23–6.

1970 'Joseph Hoare', *Jesus College Record*, 17–21.

1971 *Jesus College, Oxford, 1571–1971*, Oxford, 153 pp.

d. Other Publications

1930 'Geography', *Oxford Magazine*, Vol. 48, 673–6.

* 1932 'The climate of England in the seventeenth century', *Quarterly Journal of the Royal Meteorological Society*, Vol. 58, 421–39.

1939   (with C.F.W.R. Gullick, E.W. Gilbert and R.W. Steel) 'An atlas of the War', Oxford Pamphlets on World Affairs, No. 22. Reprinted 1940: revised edition 1941; fourth edition 1941.

1944   (with E.W. Gilbert) 'The doctrine of an axial belt of industry in England', *Geographical Journal*, Vol. 103, 49–72.

* 1947   'Geography and politics: the geographical doctrine of balance', *Transactions and Papers, Institute of British Geographers*, Vol. 13, 1–15.

* 1952   'Geography in the Essays of Elia', in *Indian Geographical Society Silver Jubilee Souvenir and N. Subrahmanyam Memorial Volume*, Madras, Free India Press, pp. 148–51.

1959   'A.G. Ogilvie and his place in British geography', in R. Miller and J.W. Watson (eds), *Geographical Essays in Memory of Alan G. Ogilvie*, Edinburgh, 1–6.

Robert W. Steel was John Rankin Professor of Geography in the University of Liverpool (1957–74) and Principal of University College Swansea, University of Wales (1974–82).

Chronology

1893   Born in Liverpool, 12 December

1913   Left Liverpool College

1913   Open Exhibition, Jesus College, Oxford, in Modern History

1914–19   Joined army, second lieutenant, King's Liverpool Regiment; war service in France and India

1917   Married Phyllis Marguerite Hancock, 8 November

1920   Distinction in Modern History

1921   Distinction in Diploma in Geography and Herbertson Memorial Prize

1921–22   Assistant in Geography, Oxford

1922   B.Litt., Oxford

1922–23   Lecturer in Geography, Bedford College, University of London

1923   Assistant to Reader in Geography, Oxford

1927   University Lecturer in Geography, Oxford

1931–32   Acting Director of School of Geography, Oxford

1933   Reader in Geography, Oxford

1935   Reader in Historical Geography, Oxford, and Lecturer in Geography, Jesus College, Oxford

1936   Secretary of Institute of British Geographers; Recorder of Section E (Geography), British Association for the Advancement of Science

1939–46	President of the Institute of British Geographers
1939	Tutor, Senior Bursar and Fellow, Jesus College, Oxford
1939–45	RAF Intelligence Service
1945	Elected to Oxford City Council
1946	Treasurer of the Hakluyt Society
1947	Retired from Readership in Historical Geography
1955	President of Section E (Geography), British Association for the Advancement of Science
1955–60	President of the Hakluyt Society
1962	Retired from Fellowship and Bursarship of Jesus College, Oxford
1963	Alderman of Oxford City Council
1964	Awarded the Victoria Medal of the Royal Geographical Society
1964	Lord Mayor of Oxford
1967	Retired from Oxford City Council
1971	Died in Oxford, 16 December

Hans Bobek

1903–1990

Elisabeth Lichtenberger

Hans Bobek, professor emeritus of geography, died on 15 February 1990 in his eighty-seventh year. The geography of the German-speaking language area lost a scientist who had actively helped to shape the discipline. As an explorer, a university teacher, and a founder and representative of institutions, he exhausted the existential opportunities available to the scholar. This was made possible by a rare talent: he was able to perceive the *gestalt* of phenomena (in the sense first described by Konrad Lorenz, cf. Festetics, 1988) both in a scientific and in an artistic–visionary way. Moreover Hans Bobek was one of the last representatives of the field of geography who succeeded in mapping so far uncharted territories by means of photogrammetric techniques. As a special honour a peak in the Sat Daghi area in Turkey was named after him.

Hans Bobek possessed a personality both fascinating and complicated. He will be remembered for his brilliant contributions to discussions, his rapid comprehension and his ability to grasp the quality of the whole and intricate processes intuitively, both in space and in the world of the mind. He will also be remembered for his impressive creativity, which was combined with an ability to associate and link seemingly heterogeneous matters. This brought ever new insights and stimuli for new research topics.

Hans Bobek had an extraordinary gift for painting and drawing. When he was lecturing, his audience was amazed at his ability to sketch an object swiftly, whether an orographic structure, a geological profile or traffic streams, on the blackboard. His artistic talents were put on a scientific basis in an inductive spatial context. Among his posthumous works there are innumerable drawings of orographic forms, especially desert phenomena, sections of cultural landscapes and the layout of fields and settlements, made especially in Iran. Later this talent was channelled towards the design of maps for the *National Atlas of Austria* – one might suppose that much of the graphic potential was sublimated in this way.

The enjoyment of graphic design was combined with enjoyment of subtleties of language. Hans Bobek was not only fluent in English and Italian, but was also able to communicate in French and Persian.

1. Education, Life and Work

Hans Bobek's private milieu may be sketched in broad strokes. He is definitely a representative of the 'much too short' twentieth century. He was born on 17 May 1903 in Klagenfurt, Austria. It was due to his father's position as a higher-echelon railway official that, during his childhood and youth, he experienced the milieu and lifestyle of the educated middle class in a multinational state.

Helene Procopovici, his first wife, came from a family of well-to-do landowners in Romania. She was his companion for forty-seven years and furthered his career. Family life included frequent contacts with her acquaintances from all parts of Europe, and in spite of numerous moves she succeeded, no doubt because of her remarkable gift for improvisation, in maintaining a generous and extremely hospitable lifestyle even in the hard times of the war and post-war periods.

Throughout the earlier stages in his career – Innsbruck, Berlin, Freiburg, Vienna – Hans Bobek's theoretical approaches diversified and the spectrum of empirical research became wider. His studies at Innsbruck University (1921–26: geography, history and social sciences) opened up two research paths: on the one hand, with a study of the city of Innsbruck and its surroundings, that of research into functional urban aspects, much later in life taken up again with research into central places; and on the other hand, under the influence of Johann Sölch, and furthered by the pleasure he took in mountaineering, research into fluviatile and glacial geomorphology in the Alps (Bobek, 1933), later continued in Iran.

At the age of 28 he was offered the position of an assistant by Norbert Krebs. Thus he had the chance to work in the then leading department of geography in the German-speaking countries, namely that in Berlin (1931–39/40). There he met a group of young scientists who were to influence the development of the discipline after the Second World War, including Carl Troll (ecology and vegetation geography), Julius Büdel (climatology and geomorphology), Herbert Louis (geomorphology) and Anneliese Krenzlin (research into settlement history) (see Lichtenberger, 1984). Krebs had expected Bobek to concentrate on urban geography; however, the latter presented his most recent research into geomorphological problems. Some of his findings contradicted the findings of Albrecht Penck, and thus met with initial disapproval. This might have been a traumatic experience, but Bobek did not give up research into quaternary problems.

Berlin's Humboldt University offered possibilities for research abroad. Hans Bobek chose the Near and Middle East, especially Iran, as 'few geographers had carried out research there before', and he found a brilliant teacher from whom he would learn Persian. Thus he fixed a lifelong regional focus for his research at the age of 30. He made the following expeditions to Iran: 1934 (eight months), 1936 (three months), 1956 (seven months, together with a limnologist, H. Löffler) and 1975 (one month: cf. *Iran. Probleme eines unterentwickelten Landes alter Kultur*).

Unlike most other scientists of his age-group, Bobek was not deprived of research undertakings during the Second World War. He was detailed to the geographical intelligence unit of the Supreme Command of the German army (1940–43). As the head of the working group for the Near and Middle East and, later on, Northern Africa too, he had access to all maps, aerial photographs and texts, whether published or unpublished, relating to these areas. During this period he wrote a paper entitled 'Soziallandschaften des Orients', a manuscript that was unfortunately never published, but formed the basis for his later social-geographical studies.

In 1944 Bobek was detailed to military research units in Russia, northern Italy and Prague; these experiences widened his geographical knowledge. By means of co-operation and discussion with ecologically oriented botanists such as Heinz Ellenberg and Josef Schmithüsen he became acquainted with geo-ecological problems, a research field he concentrated on in his studies in Iran on the natural forests and spinneys (1951) and on the climatological and ecological structure of Iran (1952).

In 1946 Bobek was appointed to a chair at Freiburg University in southern Germany (1946–48). Though the institute was in ruins, these years were very fruitful ones: he wrote pioneering papers entitled 'Die Stellung und Bedeutung der Sozialgeographie', 'Soziale Raumbildungen am Beispiel des Vorderen Orients' (1951) and 'Die Landschaft im logischen System der Geographie'.

After a short interlude at Vienna's School of Economics (1949–51), Bobek was appointed to the chair (previously held by Hugo Hassinger) at Vienna University, a position he was to hold from 1951 to 1971. It was this institutional and personal milieu that moulded the last third of his scholarly career: on the one hand he became head of the Commission on Regional Research of the Austrian Academy of Sciences (founded by Hugo Hassinger in 1946) in 1954, and on the other hand he had intensive contacts with the cartographer Erik Arnberger (Lichtenberger, 1988) as well as the planning expert Rudolf Wurzer. For the latter, Bobek drafted the very first atlas on regional planning (Lavant Valley, Province of Carinthia, Austria) and helped him in founding the Österreichische Gesellschaft zur Förderung von Landesforschung und Landesplanung after Rudolf Wurzer had been appointed to a chair at Vienna's Technical University. Moreover Wurzer succeeded in persuading Bobek to contribute to an extensive experts' report for the Austrian government, headed by Chancellor Klaus (Bobek, 1979).

It was the decision to publish a national atlas of the Republic of Austria, with the co-operation of the cartographer Erik Arnberger, that engaged the last third of his scholarly career (see Bobek, 1975). The Treaty of Vienna of 1955 provided the impetus for this project. As chairman of the Commission on Regional Research of the Austrian Academy of Sciences (1954–83), he arranged publication of the *Atlas der Republik Österreich* (*Atlas of the Republic of Austria*) and of some volumes of *Schriften der Kommission für Raumforschung* and *Beiträge zur Regionalforschung*. In the process he won the co-operation of Austria's intellectual elite from several fields who concerned themselves with spatial problems. Bobek helped prepare nearly fifty maps. Highly original and pioneering maps were published in fields as varied as climatic types, ecological evaluation, types of community, central places, and spatial economic structures. With this atlas and the publications pertinent to it, Bobek created the basis for knowledge of the spatial structure of Austria.

Another of Bobek's scholarly achievements was his research into central places, which combined regional research with regional politics. The results, presented in maps and texts, formed the background for sectional and regional planning projects of federal and provincial authorities. In this context Maria Fesl must be mentioned. She had become Hans Bobek's second wife in 1978 after he had been widowed. By then she had been his collaborator in the Commission on Regional Research for many years, and she assisted him in completing the *Atlas der Republik Österreich* and then wrote parts of the two volumes on central places (Bobek with Fesl, 1978 and 1983).

Naturally a price had to be paid for all these achievements. Many of the findings of Bobek's research, such as those derived from expeditions dedicated to geomorphological, ecological and social-geographical studies in Iran, were never published. And he never published the handbook on social geography that was advertised as forthcoming for many years by De Gruyter. Thus many of his ideas may be found only in lectures and papers, as is also true of his global insights into the spatial organization of society.

2. Scientific Ideas and Geographical Thought

Hans Bobek outlined more research themes than he could attend to. When perusing his list of publications, one finds that he had wide-ranging interests. Thus he was a cosmopolitan in thought (Bobek, 1960), and his knowledge was interdisciplinary, but he was not an encyclopaedist. He was an author of brilliant essays, among them the widely read 'The main stages in socio-economic evolution from a geographical point of view' (1962a); he did not write textbooks.

Hans Bobek's conviction as to an integrative function for geography as a science may have stemmed from his trying to cope with the heterogeneity of research topics rooted in his personality. In any case, the 'logical system of geography', developed in co-operation with Schmithüsen in the 1950s (1957, reprinted 1967), with its integrative view and ideal-typical conception of 'landscape' (Bobek with Schmithüsen, 1949, reprinted 1967), greatly influenced the discipline as a basic ideology for nearly two decades. And so another attempt had been made to mend the division of geography into a number of independent geo-branches of systematic knowledge that had already been referred to in the *Handbuch der geographischen Wissenschaft* (Klute, 1930).

Being very sure of his basic ideology, Hans Bobek was never really ready to discuss and have criticized the paradigm of the landscape he had created. Therefore the advent at the Deutscher Geographentag in Kiel in 1969 (Lichtenberger, 1978) of the new paradigm of analytical and quantitative geography formulated in the English-speaking countries, with neo-Marxist and 'analytical' undergraduates together crusading against traditional geography, must have hurt him. It is most unfortunate that he never came to terms with this break in the development of geography as a discipline, and one may assume that he no longer wished to write the book on social geography he had announced years before. To Bobek's credit he tried, at the age of sixty-five, to familiarize himself with the new techniques by attending statistics classes. He accepted, at the age of nearly seventy, the pluralism of scholarly approaches in geography; thus the analytical science theory in his farewell lecture, 'The development of geography – continuity and radical change'.

A paradox remains; this scholar, who has been regarded as a social geographer *sensu stricto*, remained a geo-scientist, both in research activities and in interdisciplinary co-operation. Highly diversified quaternary research was carried out, though intermittently, for almost four decades: two accomplishments must be mentioned. In his second thesis, for his habilitation (the qualification to become a university lecturer in German-speaking countries), on the terraces in the Inn Valley, he mapped the ice-margin traces of the period of ice decay during the late stages of the *Würm* glacial in the Alps, so that the concept of a *'Schlußeiszeit'* ('final ice age') could be dispensed with (Bobek, 1935). And in northwestern Iran Bobek was able to prove by means of field studies and the analysis of aerial photographs that during the *Würm* glacial the cold periods were also dry, whereas the climate became warmer as well as moister to the present (Bobek, 1969). Yet Bobek is not thought of as a physical geographer, but as the social geographer who developed an entirely new perspective in research (Bobek, 1948, reprinted 1969).

By developing ideas that had been formulated by Max Weber and Wilhelm Sombart and reflecting on his research in the Near East, he created, on a metasystematical level, the theory of rent capitalism (1974), without which research into underdeveloped countries (Bobek, 1962c) would be restricted. Bobek did not always translate his wealth of original ideas into formal research strategies. One might suppose that he accepted a division of labour with the Munich school of social geography (Maier *et al.*, 1977), whose inductive tradition preferred a methodological approach on the micro-level of social groups – details of which 'did not really interest [Hans Bobek] at all'.

When labelling Bobek a social geographer, one tends to overlook the fact that he had created the functional approach in urban studies and research into cities and their surroundings with his doctoral thesis (Bobek, 1928; cf. Bobek, 1927, reprinted 1969). For the first time various models of urban–rural relationships, especially the formation of monofunctional territories, were studied (Bobek, 1938, reprinted 1969), and Walter Christaller's theory of central places was augmented (Bobek, 1967a).

3. Influence and Spread of Ideas

Now that more than a quarter of a century has passed since the Kiel geographers' meeting of 1968, an attempt should be made to separate those influences Hans Bobek exercised while holding a chair from his long-term influence.

In 'traditional geography', Bobek wins recognition as the founder of the paradigm of *landschaft* in German geography, and there is general consensus that he had a marked influence on the post-war generation of geographers in the German-speaking countries. In the 1970s debate about theory, the status of the landscape paradigm as the basis for a scientific geography was rejected. In retrospect one realizes that geography thereby abandoned an important potential research area. In today's ecology-oriented society, issues concerning analysis of landscapes, landscape protection and landscape planning are of ever-increasing significance. Meanwhile neighbouring disciplines, including those of biologists and urban ecologists, have invaded this abandoned research area.

It appears that for Bobek normative societal parameters were not included in his research perspective. This is true of both landscape geography and social geography. Neither crisis phenomena of society nor problems of disparities were discussed. In retrospect this seems astonishing, and can only be understood when considering Bobek's essential position on scientific ideology: in all of his research he remained a geo-scientist and a positivist. This positivism also shaped his research into topics of human geography, and especially central places.

There is consensus that Hans Bobek, creator of social geography, fixed the framework for a new research perspective, but there are some misconceptions as to its position in the history of geography. There is no 'Vienna–Munich School of Social Geography', a term introduced in a period of heated debate about geography in the 1970s (Thomale, 1972), in order to distinguish a 'traditional social geography' from a new, micro-analytical social geography related to behavioural science as well as a 'radical geography' (Weichhart, 1993). In this context the concepts of Hans Bobek (Vienna) and Wolfgang Hartke (Munich) were lumped together, and the fundamental difference as to spatial dimensions was overlooked. The commentators missed the essential differences in scientific outlook. Whereas Bobek's social geography focused on a spatial macro-level and was concerned with cultural stages as represented in large areas, Wolfgang Hartke's main interest lay in empirical research into finding indicators for socio-geographical processes on a spatial micro-level, at the scale of individual plots and buildings. There was still another decisive difference: Bobek is considered one of the last geographers to undertake voyages of discovery, and his social geography encompasses findings made on such expeditions to the Near and Middle East, especially Iran. There he conceived the basis for his theory of rent capitalism, which he used to contrast with productive capitalism. This theory was widely adopted outside geography and became important in ethnology and elsewhere. It is a pity that Bobek did not publish the extensive empirical findings he had gathered over the decades on the organization of rent capitalism in the Near East, with many details on the complex splitting up of the production factors of land, labour and capital in a complicated structure of power, property and utilization systems.

Within his own country Bobek (as opposed to Wolfgang Hartke) did not carry out any socio-geographical research proper, and he did not develop a specific research methodology. He did not study industrial society, and left it to his research students to concern themselves with the problem of the formation of social groups (Bobek, 1973).

This lack of interest on the part of Bobek might have induced the younger generation of geographers in Vienna to take an interest in theories of the social and economic sciences and develop a research methodology combining hermeneutics and analysis. Its research topics, such as research into metropolises and into problems of the housing, labour and leisure markets, do not belong to traditional social geography in a narrow sense, but do fit in with the tasks Bobek had ascribed to social geography as a framework for human geography. Today's social geography in Austria is largely emancipated from the work of Bobek, but his influence is still pronounced in research into the Near and Middle East in the German-speaking countries, in quaternary research, in ecological research (degradation of forests), and in research into cultural history in rural areas.

As to urban geography, however, Bobek's thesis on Innsbruck created a canon of methods and a catalogue of problems that was imitated in hundreds of monographs about small and medium-sized towns for well over three decades. His early research into cities and their surroundings, as well as into functional relationships, continues to have an influence in urban geography as practised in Germany.

Finally, something should be said of Hans Bobek's influence in Austria. His predecessor, Hugo Hassinger, had studied problems of regional research and regional planning and had contributed to their advance in Austria. Bobek's attitude toward these fields remained ambivalent. Because of his liberal political outlook, he used to express scepticism as to the efficiency of any planning; on the other hand, he had stressed the importance of regional research as a topic for geography as early as his Berlin period (Bobek, 1942), and reaffirmed this opinion later in Vienna, but did not actively pursue it himself. These might be among the reasons which induced the following generation to lay stress on applied research. With the introduction of a course of studies entitled 'Regional research and regional planning' at Vienna University, the present author created the basis for a new type of professional career that has become well established in the labour market (Lichtenberger, 1980).

Bibliography and Sources

1. OBITUARIES AND REFERENCES ON HANS BOBEK

Hartke, W., 'Der Weg zur Sozialgeographie. Der wissenschaftliche Lebensweg von Professor Dr. Hans Bobek', *Mitt. d. Österr. Geogr. Ges.*, Vol. 105 (1963), 5–22.

Lichtenberger, E., Hans Bobek – ein Nachruf, *Mitt. d. Österr. Geogr. Ges.*, Vol. 132 (1990), 238–48.

Schwarz, W., 'Beiträge zur angewandten Humangeographie. Hans Bobek zum 80. Geburtstag', in *Aktuelle Beiträge zur angewandten Humangeographie: ÖIR–Forum*, Vol. 7, 5–7.

Stiglbauer, K., 'Hans Bobek und die deutsche und österreichische Landeskunde. Aus Anlaß des 80. Geburtstages am 17. Mai 1983', *Berichte zur deutschen Landeskunde*, Vol. 57, No. 1 (1983), 5–11.

Stiglbauer, K., 'Hans Bobek zum 80. Geburtstag. Bericht über das internationale Symposium "Neue Perspektiven in der Humangeographie" am 27. und 28. Oktober 1983', *Mitt. d. Österr. Geogr. Ges.*, Vol. 126 (1984), 7–10.

Wiche, K., 'Vorwort, "Festschrift Hans Bobek"', *Mitt. d. Österr. Geogr. Ges.*, Vol. 105 (1963), 3–4.

Wurzer, R., 'Zum Geleit "Festschrift zum 60. Geburtstag von Hans Bobek". Beiträge zur Raumforschung', *Schriftenreihe der Österreichischen Gesellschaft zur Förderung der Landesforschung und Landesplanung*, Vol. 2 (1964), 3–4.

a. Autobiographies
'Some comments toward a better understanding of my scholarly life-path', in A. Buttimer (ed.), *The Practice of Geography*, London–New York, Longmans, 1983, 167–85.

'Einige Bemerkungen zum besseren Verständnis meines wissenschaftlichen Lebensweges', *Geographischer Jahresbericht aus Österreich*, Vol. 48 (1991), 7–40.

b. References

Festetics, A., *Konrad Lorenz. Aus der Welt des großen Naturforschers*, München, Deutsches Taschenbuch-Verlag, 1988.

Klute, Fr. (ed.), *Handbuch der geographischen Wissenschaft.* Potsdam, Akademische Verlagsgesellschaft Atheneion. 1930, 2 vols.

Lichtenberger, E., 'Quantitative geography in the German-speaking countries', *Tijdschrift voor Economische en Sociale Geografie*, Vol. 69, No. 6 (1978), 362–73.

Lichtenberger, E., 'Zur Standortbestimmung der Universitätsgeographie. Reflexionen über die institutionelle Situation in der BRD und in Großbritannien', *Mitt. d. Österr. Geogr. Ges.*, Vol. 122, No. 1 (1980), 3–48.

Lichtenberger, E., 'The German-speaking countries', in R.J. Johnston and P. Claval (eds), *Geography since the 2nd World War. An International Survey*, London, Croom Helm, 1984, 156–84.

Lichtenberger, E., 'Erik Arnberger. Nachruf', *Almanach der Österreichischen Akademie der Wissenschaften*, Vol. 138 (1988), 412–18.

Maier, K., R. Paesler, K. Ruppert and F. Schaffer, *Sozialgeographie* in series Das Geographische Seminar, Braunschweig, Westermann, 1977.

Thomale, E., 'Sozialgeographie. Eine disziplingeschichtliche Untersuchung zur Entwicklung der Anthropogeographie', *Marburger Geographische Schriften*, Vol. 53, 1972.

Weichhart, P., 'Mikroanalytische Ansätze der Sozialgeographie – Leitlinien und Perspektiven der Entwicklung', *Innsbrucker Geographische Studien*, Vol. 20 (1993), 101–15.

2. SELECTED BIBLIOGRAPHY OF THE WORKS OF HANS BOBEK

1927   'Grundfragen der Stadtgeographie', *Geographisches Anzeiger*, Vol. 28, 213–24. Reprinted in P. Schöller (ed.), *Allgemeine Stadtgeographie. Wege der Forschung 181*, Darmstadt, Wissenschaftliche Buchgesellschaft, 1969, 195–219.

1928   'Innsbruck, eine Gebirgsstadt, ihr Lebensraum und ihre Erscheinung', *Forsch. z. Dt. Landes- u. Volkskunde*, Vol. 25(3).

1933   'Die Formenentwicklung der Zillertaler und Tuxer Alpen', *Forsch. z. Dt. Landes- u. Volkskunde*, Vol. 30.

1935   'Die jüngere Geschichte der Inntalterrasse und der Rückzug der letzten Vergletscherung im Inntal', *Jb. d. Geol. Bundesanstalt Wien*, Vol. 85, 135–89.

1938   'Über einige funktionelle Stadttypen und ihre Beziehungen zum Lande', *Comptes Rendus, Congr. Intern. Geogr., Amsterdam, 1938*, Part III, 88–102. Reprinted in P. Schöller (ed.), *Allgemeine Stadtgeographie. Wege der Forschung 181*, Darmstadt, Wissenschaftliche Buchgesellschaft, 1969, 269–87.

1942   'Geographie und Raumforschung', *Raumforschung und Raumordnung*, Vol. 6, No. 10/1, 336–42.

1948   'Soziale Raumbildungen am Beispiel des Vorderen Orients', *Tagungs-bericht Deutsches Geographentag München, 1948*, Landshut, Amt f. Landeskunde, 1951, 15 pp.

1948   'Die Stellung und Bedeutung der Sozialgeographie', *Erdkunde*, Vol. II, No. 1/3, 118–25. Reprinted in W. Storkebaum (ed.), *Sozial-geographie*, Darmstadt, Wissenschaftliche Buchgesell-schaft, 1969, pp. 44–62.

1949   with J. Schmithüsen, 'Die Landschaft im logischen System der Geographie', *Erdkunde*, Vol. III, 112–20. Reprinted in W. Storkebaum (ed.), *Zum Gegenstand und zur Methode der Geographie*, Darmstadt, Wissenschaftliche Buchgesellschaft, 1967, pp. 57–76.

1951   'Die natürlichen Wälder und Gehölzfluren Irans', *Bonner Geographische Abhandlungen*, Vol. 8, 62 pp.

1952   'Beiträge zur klimaökologischen Gliederung Irans', *Erdkunde*, Vol. VI, 65–84.

1952   'Hugo Hassinger (Nachruf)', *Almanach d. Österr. Akademie der Wissenschaften*, Vol. 102, 277–90.

1957   'Gedanken über das logische System der Geographie', *Mitt. Geogr. Ges. Wien*, Vol. 99, 122–45. Reprinted in W. Storkebaum (ed.), *Zum Gegenstand und zur Methode der Geographie*, Darmstadt, Wissenschaftliche Buchgesellschaft, 1967, 289–329.

1960   'Die spezifische Stellung und Leistung des Abendlandes', *Wissenschaft und Weltbild*, Vol. 1, 169–78.

1962a   'The main stages in socio-economic evolution from a geographical point of view', in P.L. Wagner and M.W. Mikesell (eds), *Readings in Cultural Geography*, University of Chicago Press, 218–47.

1962b   *Iran. Probleme eines unterentwickelten Landes alter Kultur*, Frankfurt–Berlin–Bonn, Moritz Diesterweg, 74 pp.

1962c   'Zur Problematik der unterentwickelten Länder', *Mitt. d. Österr. Geogr. Ges.*, Vol. 104, 1–24.

1967a   'Die Theorie der Zentralen Orte im Industriezeitalter', *Tagungsberichte und Wissenschaftliche Abhandlungen des Deutschen Geographentages Bad Godesberg 1967*, 199–213.

1967b   (ed.) *Zum Gegenstand und zur Methode der Geographie*, Darmstadt, Wissenschaftliche Buchgesellschaft.

1969   'Zur Kenntnis der südlichen Lut – Ergebnisse einer Luftbildanalyse', *Mitt. Österr. Geogr. Ges.*, Vol. 111, 155–92.

1972   'Die Entwicklung der Geographie – Kontinuität und Umbruch', *Mitt. Österr. Geogr. Ges.*, Vol. 114, 3–18.

1973   'Der Beitrag der Geographie zur Bevölkerungs- und Sozialforschung in Österreich', in *Beiträge zur Bevölkerungs- und Sozialgeschichte Österreichs*, Vienna, Verlag Geschichte und Politik, 19–28.

1974   'Zum Konzept des Rentenkapitalismus', *Tijdschrift voor Economische en Sociale Geografie*, Vol. XV, No. 2, 74–8.

1975 'Österreichs Regionalstruktur im Spiegel des Atlas der Republik Österreich', *Mitt. d. Österr. Geogr. Ges.*, Vol. 117, 117–62.

1978 with M. Fesl, *Das System der Zentralen Orte Österreichs. Eine empirische Untersuchung.* Schriften der Kommission für Raumforschung 3, Graz–Cologne, Böhlau.

1979 *Strukturanalyse des österreichischen Bundesgebietes.* 3 vols, ed. R. Wurzer: (a) *Ausgliederung der Strukturgebiete der österreichischen Wirtschaft.* (b) *Die Zentralen Orte und ihre Versorgungsbereiche. Schriftenreihe der österreichischen. Gesellschaft für Raumforschung und Raumplanung*, Vol. 2, 451–60 (a) and 475–504 (b).

1983 with M. Fesl, *Zentrale Orte Österreichs II. Beiträge zur Regionalforschung 4*, Vienna, Verlag der Österreichischen Akademie der Wissenschaften.

Prof. Dr. Elisabeth Lichtenberger holds the chair for Geography, Regional Research and Regional Planning in the Department of Geography at Vienna University, Vienna, Austria.

Chronology

1903 Born in Klagenfurt, Austria, 17 May

1921–26 Student at Innsbruck University; subjects: geography, history, social sciences

1931–39/40 Assistant at the Department of Geography, Humboldt University in Berlin, with Norbert Krebs

1940–43 Military geographical research unit of the Supreme Command Department of the German army

1944–45 Military research unit in Russia, Yugoslavia, northern Italy, Prague

1946–48 Chair at Department of Geography, Freiburg University, Federal Republic of Germany

1949–51 Chair at Department of Economic Geography, Vienna School of Economics

1951–71 Chair at Department of Geography, Vienna University; chairman of this department

1952 Visiting Professor, Department of Geography of the University of Nebraska at Lincoln, USA (summer term)

1954–69 Proponent for the foundation of the Österreichische Gesellschaft zur Förderung von Landesforschung und Landesplanung, and its vice-chairman

1954–83 Chairman of the Commission on Regional Research of the Austrian Academy of Sciences

1958–59 Visiting Professor, Tehran University, Iran (academic year 1958/59)

1963–64	Visiting Professor, Technical University Munich, Federal Republic of Germany (academic year 1963/64)
1973	Visiting Professor at various universities in Japan (3 months)
1978	Honorary Ph.D. of Bochum University, Federal Republic of Germany
1990	Died in Vienna, Austria, 15 February

Franciszek Bujak

1875–1953

Aldona Ertman

Franciszek Bujak was educated in history and became a keen enthusiast of geography. He joined both disciplines together in his research and monographic works, and he became interested in the history of geographical science. His contribution warrants attention as that interdisciplinary posture reflected the scientific tendency of that period in Polish intellectual history, and also because of his role in the emergence of the historical and geographical sciences.

1. Education, Life and Work

Franciszek Bujak was born in 1875, the son of a farmer in the village of Maszkienice, near Cracow. His education in the village and at Bochnia took place at a time when the educational system of Galicia was undergoing reform and improvement. It was decisive for his thorough educational preparation, his patriotism, and for a lifelong commitment to learning.

Contrary to the then prevailing tendency to educate peasants' sons for the clergy, Bujak – despite the opposition of the family – chose, in 1894, to undertake studies in the Faculty of Philosophy at the Jagellonian University in Cracow. Historians influenced his research interest: Wincenty Zakrzewski in the field of economic history, Stanisław Smolka in the field of source studies on the history of Poland, and B. Ulanowski, with whom he collated historical and juristical sources for the history of Poland. The latter studies were related to palaeography. He developed an interest in anthropology, but ethnography and geography interested him most.

Bujak was encouraged in his geographical studies by the Chairman of Comparative Physical Geography at the Jagellonian University (the chair had been inaugurated in 1849, was active to 1853, and was re-established in 1877). The chair was held by Franciszek Szwarcenberg-Czerny, a scholar of comprehensive geographical interest, including the history of geography, who was familiar with scientific developments in Austria and Germany.

Bujak's historical and geographical studies resulted in a dissertation entitled 'The Development of Geography in Poland in the Middle Ages and at the Beginning of the Sixteenth Century'; this study led to the award of the degree of Doctor of Philosophy in 1899.

Bujak began his professional work as a librarian at the Jagellonian University. When he was awarded a scholarship for 1901–02, he travelled to Leipzig and studied with F. Ratzel, K. Bücher, K. Lamprecht, H. Berger and E. Kötzschke. That experience allowed him to transfer to Poland the research trends he had observed in Germany. During visits in 1903 to Genoa, Florence, Rome and Munich, Bujak also compiled source materials for his study on medieval and nautical maps.

Curiously, while Bujak was involved in numerous radical liberal activities, he was sponsored by Szwarcenberg-Czerny, who was a pro-Austrian monarchist and an adversary of the national liberation movement of the Polish peasantry. That is why Bujak's post-doctoral dissertation 'Studies of Settlement in Little Poland' (1905) dealt with economic history and corresponded to the anthropogeographic themes of Ratzel's geographical seminar.

In 1909, Bujak was nominated Extraordinary Professor in Economic and Social History at the Jagellonian University: he remained there until 1919. After a one-year stay in Warsaw, he moved in 1921 to the University of Lvov as an ordinary professor in Polish and universal economic history. He held that position up to the outbreak of the Second World War, at which time he went into hiding. In 1945, Bujak returned to Cracow and the Jagellonian University, where he retained a professorship until retirement. He died on 21 March 1953.

In addition to his pedagogic and scientific activities at the Jagellonian University, Bujak held offices in the Academy and in other scientific institutions in Cracow, and was always active and influential. Bujak's scientific publications total some four hundred items. Many of them are the fruit of the first period of his close association with the mother-university, which lasted more than ten years; during this time nearly sixty items were produced in the fields of the history of geography and historical geography. Most numerous are his contributions to economic and social history, being of substantial importance for geography, and particularly for historical geography. Bujak was also interested in problems of science, cultural institutions and culture itself. The economic problems of Galicia and ethnography were also part of his concern.

2. Scientific Ideas and Geographical Thought

Anthropogeography, the history of geography and historical geography were, from the beginning, the fields of Bujak's scientific activity. This was in accord with the tradition at Cracow.

While K. Miller was systematizing and publishing his 'mapemundi' (1898), Bujak carried on his source studies at the Jagellonian University for his Ph.D. dissertation concerning Polish geography in the Middle Ages. He studied the maps drawn and elaborated in Cracow, and wrote on the development of cartography in one of the many European geographical centres. He indicated that the medieval codices of Cracow reflected the cosmographic thought of western Europe, whereas in the second half of the sixteenth century the works of some scholars suggested the full acceptance of Ptolemaic geography and original application of the Ptolemaic cartographical method. He presented the achievements of the scholars of the Jagellonian University, known in the European literature, and stressed their merits for Polish geography: Jan Długosz (*Chorografia Regni Poloniae, c.* 1480), Johanes of Glogow (*Introductorium Cosmographiae*, Ulm, 1486), Laurencius Corwinus (*Cosmographia dans* [*sic*] *Manuductionem in Tabulas Ptholomei*, Basel, 1496). He also presented the historiographers who contributed to geography: Jan Długosz and Filippo Kallimach (Buonacorsi). In this way, the history of science as studied at the Jagellonian University was joined with European geography, providing evidence, Bujak thought, for the coherence of the Christian culture of the Middle Ages and the Renaissance.

As a result of his visits to other countries and his study of the literature, Bujak wanted to strengthen the standing of both the history of geography and historical geography. The studies by Birkenmajer in *Marco Beneventano, Kopernik, Wapowski, a najstarsza karta geologiczna Polski* (1901) (*Marco Beneventano, Copernicus, Wapowski and the oldest geological map of Poland*) and the index of sources by Lewicki, 'Codex epistolaris saec. XV' (1891) formed the basis for Bujak's 'Początki kartografii w Polsce' (1900) ('The beginnings of cartography in Poland'). That work in turn provided inspiration and opportunity for further research in the emergence of geographical science.

Bujak's study 'O średniowiecznych mapach żeglarskich' (1903) ('On medieval nautical maps') encouraged the development of a Polish interest in nautical cartography which was already developed in other parts of Europe (H. Wagner, 1895, 1896, 1900; E. Steger, 1896; A.F. Nordenskiöld, 1897). He based his studies on the little-known manuscript 'Nota ad chartam navigandi' (1490), by Burchard de Anvil, which was an introduction to nautical maps. He thus shed light upon the origin and methods of constructing navigational instruments, and upon some of the essential problems of nautical cartography.

The work accomplished by K. Lamprecht and E. Kötzschke at Leipzig, expanded by additional studies, was valuable to Bujak in the preparation of his study of the historical maps of the Polish kingdom in 'W sprawie kartografii historycznej' (1906) ('On historical cartography'). His lectures and publications on the historical geography of Poland were later of significance to W. Semkowicz (1922).

Perhaps of particular significance was Bujak's interest in the location of the capital of Poland, which he wrote about in physical and political contexts in several of his publications. He was also interested in Galicia – its social demands and aspirations for political and economic emancipation. His comprehensive monographs of three towns – Maszkienice (1901), Limanowa (1902) and Żmiąca (1903) – were followed by 'Studia nad osadnictwem Małopolski' (1905) ('Studies on the settlement of Little Poland').

Bujak's most comprehensive geographic-economic work was *Galicja* (1908 and 1910) (*Galicia*) – a comprehensive analysis of the economy of that Polish region. He continued his research concerning this region and later published his findings in two volumes: *Z odległej i bliskiej przeszłości Studia historyczne-gospodarcze* (1924) (*On the Distant and Recent Past. Historical-Economic Studies*) and *Studia historyczne-społeczne* (1924) (*Historical and Social Studies*).

3. Influence and Spread of Ideas

Bujak's scientific activity began at a time when disciplines were being formed and influenced by ideas of positivism. The names of emerging sectors of the discipline were provided by the titles of the compendia devoted to them. Examples include *Anthropogeographie* (1882) and *Politische Geographie* (1891), both written by Friedrich Ratzel.

The studies on Polish geography inspired by Bujak were continued on a larger scale by Władysław Semkowicz in *Rozwój nauk pomocniczych historii w latach 1886–1936* (1937) (*Development of Sciences Auxiliary to History in the Years 1886–1936*) and *Rozwój nauk pomocniczych historii w Polsce* (1958) (*Development of Sciences Auxiliary to History in Poland*) and by Tadeusz Ładogórski in *Atlas historyczny Polski* (1976) (*Historical Atlas of Poland*). Julian Janczak and Tadeusz Ładogórski were editors of these works.

Stefan Inglot, author of *Zarys dziejów nauk rolniczych i leśnych w Polsce* (1948) (*An Outline of Agricultural and Forest Sciences in Poland*) and of *Zarys polskiej myśli ekonomiczno-rolniczej po II Wojnie Światowej* (1973) (*An Outline of Polish Economic-agricultural Thought after World War II*) continued Bujak's research tradition with regard to agricultural geography and its history. His work in the field of economic history was further considered and extended by a number of scholars including Karol Maleczyński, Jan Rutkowski and Stanisław Pazyra. Bujak's studies on the history of cartography were continued by Karol Buczek, the author of *History of Polish Cartography* (1982).

Bujak was the editor of a series of publications which initiated later research trends. In the series *Badania z dziejów społecznych i gospodarczych* (*Studies on Social and Economic History*), 34 volumes of which were published commencing in 1925, social problems and the outbreak of plagues in Poland and in the neighbouring countries in 1450–1586 were presented. In the series *Biblioteka Puławska* (*Puławy Library*), which yielded 19 volumes in the years 1927–31, monographs of several villages were published, which helped develop the subject of rural geography. Two other series, namely *Zadania i potrzeby gospodarcze* (1916–1917) (*Economic Problems and Tasks*) and *Roczniki dziejów społecznych i gospodarczych* (1931–39 and 1946–51) (*Annals of Social and Economic History*), concentrated on economic history and its geographic conditions.

At the beginning of this century, Bujak infused Polish geography with the ideas and research trends developed in Germany and other European countries; as a result of his initiative Polish geography was strengthened as it entered the twentieth century.

Bibliography and Sources

1. REFERENCES ON FRANCISZEK BUJAK

Gąsiorowska, Natalia, 'Franciszek Bujak, 1875–1953. Obituaries', *Nauka Polska*, No. 3 (1953), 185–8.

Inglot, Stefan, 'Franciszek Bujak. Działalność naukowa i pedagogiczna' ['Franciszek Bujak, scientific and didactic activity'], *Sprawozdanie Wrocławskiego Towarzystwa Naukowego*, No. 8 (1953), Appendix 3.

Kaczmarczyk, Kazimierz, 'Franciszek Bujak, 1875–1953', *Przegląd Zachodni*, No. 4/5 (1953), 706–12.

Kozłowska-Budkowa, Zofia, 'Działalność naukowa Franciszka Bujaka' ['Scientific activity of Franciszek Bujak'], *Studia Historyczne Yearbook*, Vol. 1, No. 3/4 (1958), 5–18.

Madurowicz-Urbańska, Helena, *Franciszek Bujak Wybór pism* [*Selected Works*], Vols 1 and 2, Warsaw, 1976.

Sródka, Andrzej, and Paweł Szczawiński, 'Biogramy uczonych polskich' ['Biograms of Polish scholars'], *Nauki społeczne*, Part 1, fasc. 1: A–J.

Szczotka, Stanisław, 'Professor Dr Franciszek Bujak', *Lud*, Vol. XLV (1954), Part 1, 7–26.

2. SELECT BIBLIOGRAPHY OF WORKS BY FRANCISZEK BUJAK

History of Geography

1896 'Pamiątki po J. Lelewelu w Bibliotece Jagiellońskiej w Krakowie' ['Memorials of J. Lelewel in the Jagellonian Library in Cracow'], *Nowa Reforma*, No. 207, 2.

1896 'O ludach Cejlonu' ['On the peoples of Ceylon'], *Wisła*, Vol. 10, 645–53.

1898 'Franciszek Polacky i jego znaczenie dla odrodzenia Czech' ['Franciszek Polacky and his significance for the renaissance of Czech'], *Ateneum*, Vol. 3, 397–437.

1899 'Rozwój geografii w Polsce w średnich wiekach i na początku XVI wieku' [Development of geography in Poland in the Middle Ages and at the beginning of the XVIth century'], dissertation, 1900, part editions 1901, full edition 1925, Warsaw, 10–47.

1900 'Geografia na UJ do połowy XVI wieku' ['Geography at the Jagellonian University until the middle of the XVIth century'], in *Księga pamiątkowa uczniów Uniwersytetu Jagiellońskiego*, Cracow, 49–96.

1900 'Początki kartografii w Polsce' ['Beginnings of cartography in Poland'], *Wiadom. Numizm.-archeol.*, Annual 12, Vol. 4, No. 2, 180–6.

1900 'Najstarszy opis Ziemi Świętej polskiego pochodzenia' ['The oldest description of the Holy Land in Poland'], *Wiadom. Numizm.-archeol.*, Annual 12, Vol. 4, No. 3–4, 245–54.

1900 'Kallimach i jego znajomość państwa tureckiego w Polsce około początku XVI wieku' ['Kallimach and his knowledge of the Turkish state in Poland about the beginning of the XVIth century'], *Rozpr. Akadem. Umiej. Wydz. hist-filozof.*, 2nd series, Vol. 15, General collection Vol. 40, 268–88.

1901 'Długosz jako geograf' ['Długosz as geographer'], in *Przewodnik naukowy i literacki*, Lvov, 171–84.

1901 'Maszkienice. Wieś powiatu brzeskiego' ['Maszkienice, a village in Brzesko country'], *Rozpr. Akadem. Umiej. Wydz. hist.-filozof.*, 2nd series, Vol. 16, General collection Vol. 41, 76–184.

1902   'Limanowa. Miasteczko powiatowe w zachodniej Galicji. Stan społeczny i gospodarczy' ['Limanowa, a country town in western Galicia. Its social and economic conditions'] *Kraków, Studya ekonomiczno-społeczne*, fasc.1, viii + 221 pp.

1903   'O średniowiecznych mapach żeglarskich' ['On medieval nautical maps'], *Rozpr. Akadem. Umiej., Wydz. hist.-filozof.*, 2nd series, Vol. 21, General collection Vol. 46, 190–264.

1903   'Żmiąca. Wieś powiatu limanowskiego. Stosunki gospodarczo-społeczne' ['Żmiąca, a village in Limanowa country. Economic and social conditions'], *Kraków, Studya ekonomiczno-społeczne*, fasc. 3, vii + 152 pp.

1905   'Studia nad osadnictwem Małopolski' ['Studies of the settlement in Little Poland'], *Rozpr. Akadem. Umiej. Wydz. hist-filozof.*, 2nd series, Vol. 22, General collection Vol. 47, 172–438.

1906   'W sprawie kartografii historycznej' ['On historical cartography'], *Kwart. Hist.*, Vol. 20, 483–97.

1908   *Galicja, t.I. Kraj, ludność, społeczeństwo [Galicia, Vol. I. Land, Population, Society, Agriculture]*, Lvov, Wiedza i Życie, Ser. 4, Vol. 2, 562 pp.

1910   *Galicja t.II. Leśnictwo, Górnictwo i Przemysł [Galicia, Vol. II. Forestry, Mining, Industry]*, Lvov, Wiedza i Życie, Ser. 4, Vol. 7, vi + 509 pp.

1920   *Historia osadnictwa ziem polskich [History of the Settlement of the Polish Lands]*, Warsaw, Wydawnictwo Głównego Urzędu Ziemskiego, No. 1, 62+1 pp.

1924   'Stolice Polski' ['Capitals of Poland'], *Przegląd Wszechpolski*, Vol. 3, No. 8, 632–3.

1925   *Studia geograficzno-historyczne [Geographical-historical Studies]*, Warsaw, xi + 1 + 299 + 1 pp.

1926   'Jan Długosz, geograf polski XV wieku' ['Jan Długosz, Polish geographer of the XVth century'], *Prace geogr.*, fasc. 5, 81–126.

1927   'Drogi mojego rozwoju umysłowego' ['The ways of my intellectual development'], *Nauka Polska*, Vol. 6, 77–136.

1934   'Dwa ustępy z geografii historycznej Polski' ['Two paragraphs on the historical geography of Poland'], Lvov, 569–79.

Other Publications
1906   'Historia stosunków gospodarczych' ['History of economic relations'], *Ekonomista*, Annual 6, Vol. 2, 1–11.

1916   *Seria: Zadania i potrzeby gospodarcze [Series: Economic Problems and Tasks]*, 1916–17, 13 works.

1916   'Austriacka polityka handlowa względem Galicji w latach 1772–1790' ['Austrian trade policy towards Galicia 1772–1790'], *Kwart. Hist.*, Vol. 30, 343–56.

1918   *O naprawie ustroju rolnego w Polsce [On the Improvement of the Agricultural System in Poland]*, Warsaw, 160 pp.

1920 *O podziale ziemi i reformi rolnej* [*On Land Division and Agrarian Reform*], Wyd. III rozszerzone, 104 pp.

1921 *Uwagi krytyczne o naszej reformie rolnej* [*Critical Remarks on our Agrarian Reform*], Warsaw, iv + 107 pp.

1923 'Zagadnienie syntezy w historii' ['The problem of synthesis in history'], *Kwart. Hist.*, Vol. 37, 1–23.

1924 *Studia historyczne i społeczne* [*Historical and Social Studies*], Lvov, xi + 261 + 3 pp.

1924 'Traktat Kopernika o monecie' ['Copernicus' treatise on the coin'], in *Mikołaj Kopernik*, Lvov, 40–103.

1924 *Z odległej i bliskiej przeszłości. Studia historyczne-gospodarcze* [*On the Distant and Recent Past. Historical-Economical Studies*], Lvov, viii + 325 + 3 pp.

1925 *Badania z dziejów społecznych i gospodarczych* [*Studies on Social and Economic History*], series, 1925–48, 34 vols.

1927 *Seria pt. Biblioteka Puławska* [*Pulawy Library Series*], 1927–31, 19 vols.

1930 *Nauka o społeczeństwie* [*Science about Society*], Warsaw, 200 + 7 pp.

1931 *Roczniki dziejów społecznych i gospodarczych* [*Annals of Social and Economic History*], series, 1931–39, 1946–51.

1949 'Państwo i naród polski w XI wieku' ['Polish state and society in the XIth century'], in *Studia z Dziejów Kultury Polskiej*, Warsaw, 21–37.

3. SOURCES AND ARCHIVAL MATERIALS

de Anvil, Burchard, 'Nota ad chartam navigandi' (1490). MS. National-bibliothek, Munich.

Birkenmajer, Ludwik-Antoni, *Marco Beneventano, Kopernik, Wapowski, a najstarsza karta geologiczna Polski* [*Marco Beneventano, Copernicus, Wapowski and the Oldest Geological Map of Poland*], 1901.

Buczek, Karol, *History of Polish Cartography*, 3rd edition, Amsterdam, 1982.

Corwinus, Laurencius, *Cosmographia dans* [*sic*] *Manuductionem in Tabulas Ptholomei*, Basel, 1496.

Długosz, Jan, *Chorografia Regni Poloniae*.

Inglot, Stefan, *Zarys dziejów nauk rolniczych i leśnych w Polsce* [*An Outline of Agricultural and Forest Sciences in Poland*], Kraków, 1948, 52.

Inglot, Stefan, *Zarys polskiej myśli ekonomiczno-rolniczej po II Wojnie Światowej* [*An Outline of Polish Economic-agricultural Thought after World War II*], Wrocław, 1973.

Johanes of Glogow, *Introductorium Cosmographiae*, Ulm, 1486.

Krzymowska, Alicja, 'Franciszek Szwarcenberg-Czerny profesor geografii Uniwersytetu Jagiellońskiego, 1847–1917' ['Franciszek Szwarcenberg-Czerny Professor of Geography, Jagellonian University, 1847–1917'], *Prace Geograficzne*, No. 3, IG PAN, p.69.

Lewicki, Anatol, 'Codex epistolaris saec. XV', Cracow, 1891.

Ładogórski, Tadeusz, *Atlas historyczny Polski* [*Historical Atlas of Poland*] Wrocław, 1976.

Miller, Konrad, *Die ältesten Weltkarten*, Stuttgart, 1898.

Semkowicz, Władysław, *Rozwój nauk pomocniczych historii w latach 1886–1936* [*Development of Sciences Auxiliary to History in the Years 1886–1936*], Lvov, 1937.

Semkowicz, Władysław, *Rozwój nauk pomocniczych historii w Polsce* [*Development of Sciences Auxiliary to History in Poland*], Cracow, 1958.

Aldona Ertman is a senior assistant in the Department of the History of National Sciences in the Institute of the History of the Polish Academy of Sciences, in Warsaw.

Chronology

1875	Born in village of Maszkienice, 16 August
1894–1900	Studied at Jagellonian University (geography/history/law)
1899	Awarded a doctorate for 'The Development of Geography in Poland in the Middle Ages and at the Beginning of the Sixteenth Century'
1903	Studied in the archives and libraries of Genoa, Florence, Rome and Munich
1905	Post-doctoral habilitation, Jagellonian University; published 'Studia nad Osadnictwem Małopolski' (Studies on the Settlement of Little Poland)
1908, 1910	*Galicia*, Vols I and II
1909–18	Appointed Extraordinary Professor of Economic and Social History at Jagellonian University
1919–20	Professor, Head and Chair of Economic History at Warsaw University
1920	June–July, Minister of Agriculture and State Properties; published *History of the Settlement of the Polish Lands.*
1921–39	Professor, Head and Chair of Department of Socio-economic History (later became Institute) at J. Kazimierz University, Lvov.
1924	Published 'Capitals of Poland'
1925	Published *Geographical-historical Studies*
1933–37	President of Polish Scientific Society in Lvov.
1953	Died in Cracow, 21 March.

Xu Hongzu

1587–1641

Yu Xixian and Sha Luyin

Xu Hongzu, styled Zhenzhi, was also named Xu Xiake. He was a native of Jiangyin, Nanzhili (presently Jiangin County, Jiangsu Province), and lived during the Ming Dynasty. At the age of 20 he embarked on geographical explorations – an undertaking which he was to continue during the next thirty years or so. In his explorations he covered the greater part of China, tirelessly probing into the mysteries of nature. The places he explored were often desolate: wild mountain regions where even fishermen and woodsmen seldom went, or deep forests where the flora and fauna were various.

He travelled during the day and at night he would write about his travels by the dim light of an oil lamp. Every night he would faithfully record the bio-landscape and the local conditions and customs of the places through which he had travelled that day; he would also record his own observations.

He has left us a precious, 600,000-word, multi-volume work – *Xu Xiake's Travels*. It gives a vivid description of the natural landscapes and social history of China three hundred to four hundred years ago. His book makes it clear that he initiated a new orientation in the study of geography, namely systematically observing natural phenomena and describing geographical conditions. Basing his analyses upon simple yet precise measurements and comprehensive descriptions of geographical conditions, he investigated the relationships between geographical phenomena and made a comparative study of the geographical features of different regions. Thus he broke through the limitations imposed by classical geography and promoted development towards modern geography.

1. Education, Life and Work

As early as the Song Dynasty (eleventh to thirteenth centuries AD), Jiangyin, Xu's birthplace, was already a famous trade port. In the mid-Ming Dynasty (fifteenth to seventeenth centuries AD), a commodity economy was flourishing in Jiangyin, and here, in this place noted for its textile industry and trade, Xu Hongzu was born into a scholarly family. The family's financial situation worsened when his father's generation began to incur the expenses of a large house; it improved, however, when his mother encouraged the family to produce textiles and cultivate vegetables.

His mother was full of curiosity about the landscape in various parts of the country. This made a deep impression on the young Xu Hongzu. In later years, each time he returned home after a long journey he would recount to his mother his experiences and the new things he had seen and heard during his travels. His mother would listen to his stories with keen interest, and would sometimes raise intriguing questions which stimulated him to observe more closely and describe more accurately the geographical conditions and features of different places. His mother was the first appreciative friend he had in his career as a geographer.

Xu Hongzu's father was fond of gardening. He disliked conventional social activities and preferred to devote himself to arranging rockeries and ponds, growing flowers and other plants, and building pavilions.

During his childhood, Xu was a close observer of nature: he was fond of travelling and he became familiar with the names of many kinds of animals and plants. At the age of fourteen Xu began to read books on history and geography. Travel accounts and local chronicles fascinated him most and within a few years he was known as a very learned scholar.

When Xu was seventeen his father died, after being bedridden for some time. Xu decided to devote himself to investigating the geography of China, instead of following the career in officialdom which would have brought him literary honours, official rank and a handsome salary. Throughout his life he persisted in his chosen course, striving to obtain a genuine knowledge during his thirty years of exploration.

2. Scientific Ideas and Geographical Thought

On the basis of field investigations, Xu described in detail the geographical conditions of the Yunnan–Guizhou Plateau and its Five Ridges (spanning the borders on one side between Henan and Jiangxi and on the other between Guangdong and Guangxi), correcting the mistaken views and vague conceptions previously held about this area. He was one of the first in the world to discover the scientific concept of the watershed. He also correctly estimated the ratio between the drainage area of the Yangtze River and the Huanghe River (Yellow River) as 2:1. This was one of his major contributions to geography. He independently observed and recorded the erosion and sedimentation of water currents. In addition he divided erosion into different forms, such as vertical and horizontal erosion.

Making use of the methodology of comparative study popular in modern geography, Xu found that of two rivers with the same datum plane, given that the altitude of their sources is the same, the one whose source is a shorter distance away from the plane and which has a greater speed of flow will cause greater erosion. This scientific conclusion was based upon precise measurements of altitude and distance.

The area of karst topography in southern China occupies approximately one third of the country. Over three hundred and fifty years ago, Xu Hongzu made the first comprehensive, systematic and scientific investigation of karst topography. He classified and named these topographic features describing the characteristics of each category. The categories include: dints and karrens; karst fissures; sink holes; karst funnels; karst depressions; karst basins; karst shafts; blind valleys; dry valleys; karst bridges; and karst caves. He also observed the differences between the karst topography in different areas and determined the places in which each was found. He studied the forms and configurations of the karst topography in different regions and determined their respective distribution and demarcation lines. In Europe it was not until the 1830s that the French geographers Virlet (1839) and Th. Lyll (1839) began to investigate karst. This was between 130 and 200 years after Xu's studies on karst topography.

Xu Hongzu explored more than three hundred caves, possibly exceeding any other geographer in the number and types of caves he investigated. He studied the geographical location of each cave, the depth below the earth's surface, the configurations, the mechanisms of corrosion, erosion and collapse, the hydrographical conditions, the airflow and wind-speed, and the secondary sediments – their types, form and causes, and so forth. In short, his studies were thoroughly comprehensive, his description vivid and his interpretations scientific. He recorded position, depth, size and breadth of the chambers and the paths in these chambers as an interior landscape. These calculations are remarkably similar to the measurements taken with scientific instruments. In Europe it was not until the nineteenth century that Ivan (1771–1809) and Boris, two Russian geographers, described the caves in the Urals.

Xu Hongzu was the first in the world to formulate the theory that stalactites and stalagmites are formed by the evaporation-sedimentation of drops of water. More than a hundred years after he put forward this theory, the Russian geographer Vladimir (in his book *On Strata*, published in 1763) published a similar theory. Xu also studied various mountain areas, analysing the relationship between altitude, climate and vegetation. He pointed out that as the climate increasingly cools with latitude, the vegetation in mountainous areas changes, from evergreen broadleaf forest to deciduous broadleaf forests, from deciduous forest to mixed coniferous and broadleaf forests, from mixed coniferous and broadleaf forests to coniferous forests, from coniferous forest to bushes, and from bushes to naked ground with no vegetation. He was aware that the temperature becomes lower as one moves further north. He also observed the ecology of mountain vegetation, noting that plants growing in strong sunshine have thick stems, small joint nodes and many branches, the most common plants often being creepers and vines.

In addition, he noted the difference in temperature and degree of humidity at the same time of year between mountain areas, river valleys and plains, and thus ascertained the influence of topography on climate. He also noticed that precipitation is relatively high on the windward slope of a mountain, whereas on the leeward slope there are dry, hot winds and the precipitation is very low.

In his diaries he also recorded in detail the weather conditions on the Yunnan Plateau during the period from September 1638 to January 1641. His diaries show that the annual rainfall in the Yunnan Plateau during that time was approximately double the present annual rainfall. He also recorded that in the winter of 1616, China's famous scenic area – the Yellow Mountains (Huangshan) – was snow-clad for at least three and a half months. By contrast, nowadays the Yellow Mountains are rarely covered with snow in winter. His diaries contain descriptions of various local products and other such information, which are the concerns of human geography.

In summary, Xu Xiake's achivements rank him among the world's pioneers in modern geography.

3. Influence and Spread of Ideas

Xu's thought is reflected in his diaries. He believed that the various geographical phenomena influence and condition one another, that every region has its own distinctive geographical features, and that in order to understand these features, one has to go to different places, and collect and analyse relevant data.

After Xu died, an Italian geographer named P. Martini SJ borrowed from Xu's travelogues, compiling *Novus Atlas Sinensis* (*New Atlas of China*), printed in Amsterdam. The atlas includes 17 maps and 171 pages of written description.

Xu Xiake's (Hongzu) Travels is a book of great value to the study of geography; yet its significance was not fully recognized for a long time. Around 1940, however, a group of Chinese scholars, who returned to China after studying abroad, finally paid tribute to its importance. Since then, this book and its contribution to the development of modern geography have been the subject of much scholarly interest.

Bibliography and Sources

Xu Xiake's Travels (collated by Chu Shaotang and Wu Yingshou), Shanghai, Shanghai Ancient Books Press, 1980.

Xu Xiake's Travels (collated by Zhu Huirong), Kunming, People's Publishing House of Yunnan, 1985.

Hou Renzhi, *On Xu Xiake*, Beijing, Zhonghua Publishing House, 1961.

Hou Renzhi, 'On Xu Xiake and his travels', *The Field of Social Sciences*, Jilin, 1980.

Liu Guocheng, *Critical Biography of Xu Xiake*, Haerbin, Northeastern Forestry University Press, 1986.

Tang Xiren and Yang Wenheng, *A Study of Xi Xiake and His Travels*, Beijing, China Social Sciences Press, 1987.

Yu Xixian, 'Xu Xiake, a great geographer of the Ming Dynasty', *Journal of the Dialectics of Nature*, Beijing, 1983.

Yu Xixian, 'The significance and motives of Xu Xiake's devoting himself to geographical explorations', *Geographical Knowledge*, Beijing, No. 7, 1983.

Yu Xixian, 'Geographer Xu Xiake's contributions to the history of prose writing', *Journal of Yunnan Teachers University*, Kunming, No. 4, 1986.

Yu Xixian, *Xu Xiake – A Geographer of the Ming Dynasty*, Beijing, Popular Science Press, 1987.

Yu Xixian is a professor of geography at Beijing University, China, who has published much on Xu Hongzu.

Chronology

1587	Born in Jiangyin on 15 January
1601	Collected and read books about history, geography and travel; began to contemplate geographical exploration
1604	Decided to devote himself to geographical exploration
1607	Visited the Taihu Lake and other historic spots in Jiangsu Province, thus starting his career as an explorer-geographer
1609	Climbed Mount Tai; visited the birthplace of Confucius in Qufu (Shandong) and the birthplace of Mencius in Zouxian County (Shandong); climbed the Yi Mountain
1613	Visited Luojia Mountain via Zhejiang Province; made his first trip to the Yandang and Tiantai Mountains
1614	Visited Nanjing, winter
1616	Visited the Baiyue and Yellow Mountain in Anhui Province, and the Wuyi Mountains in Fujian Province
1617	Made a trip to the Shanjuan and Zhanggong Caves in Yixing, Jiangsu Province
1618	Visited the Lushan Mountains in Jianxi Province; made another trip to the Yellow Mountains
1620	Visited the Jiuli Lake in Xianyou Country, Fujian Province
1623	Visited the Songshan Mountains in Henan Province, the Huashan Mountains in Shanxi Province, and the Wudang Mountains in Hubei Province
1628	Made his third trip to Fujian Province and climbed the Luo-fu Mountains
1629	Made a trip north to Beijing and the Panshan Mountains in Tiajin
1630	Visited Fujian for the fourth time
1632	Visited the Tiantai and Yandang Mountains again
1633	Made his second trip to Beijing; went to the Hengshan and Wutai Mountains in Shanxi

1636–40   Made his arduous 'Ten-thousand-li Trip' to the southwestern part of China, September 1636 to June 1640. During this trip, he travelled remote and backward areas inhabited by minorities, and explored the sources of the Yangtze River and Pearl River. He passed through Jiangsu, Jiangxi, Hunan, Guangxi, Guizhou and Yunnan, reaching Zhongdian of Lijiang (Yunnan) and Tengchong, the border area between Yunnan and Burma. His achievements were chiefly made during this exploration.

1641   Died, February

Chong-ho Kim

c. 1804–1866

Hong-key Yoon

1. Education, Life and Work

Chong-ho Kim, the most famous geographer that Korea has ever produced, is believed to have lived from 1804 to 1866. Although he died less than a hundred and thirty years ago, his life is little known and not well documented. There are two important reasons why it is difficult to trace his life accurately. First, Kim did not leave autobiographical notes of any kind, nor can any reference to him be found in the genealogical record of the Kim clan of Chongdo, to which his family seems to have belonged. Secondly, there are virtually no other nineteenth-century records of his life, except for comments on him and his work made by scholars of his time.

All Korean families of the scholar gentry or *Yangban* class had a book of their own family genealogy, through which a member of the group should be able to trace his own genealogical record. The class of commoners did not usually have a family genealogy book. The fact that no record of Kim can be found in the genealogical book of the Kim clan suggests that he was a commoner. Some scholars argue that his family may have been of the middle technocrat class (*chung-in*), since he was a learned person, unlike most commoners of his time (Yi Sangtae, 1991, p. 207). However, it is difficult to accept this view, as the middle technocrat class consisted of professionals such as medical doctors, translators, geomancers and petty local magistrate officers. All evidence suggests that Kim did not belong to any one of the middle-class professions. Upward social mobility from a commoner's status to the scholar gentry or the *Yangban* class through hard work and scholastic excellence was possible, but such movement was not possible for a middle technocrat whose profession was clearly defined by his special skill. Kim was most probably a commoner with excellent scholastic abilities, not an upper-class scholar gentleman or a middle-class technocrat.

The three recently discovered non-genealogical sources which comment on Kim's life reveal little of his family origin or social status (Yi Sangtae, 1990,

p. 5). They merely comment briefly on him as a geographer and cartographer. Although his life has not been well documented, we know that Kim was from Hwanghae province (presently part of North Korea), and later moved to Seoul, where he carried out his scholarly works.

We do not know why and when he moved to Seoul, for there is no record on this matter. However, one can imagine that a young man with talent and ambition went to Seoul, the capital city, to study and make his name famous: as a Korean proverb says, 'When a boy is born, send him to Seoul [the capital city], and a colt to Cheju Island [the most famous horse-raising place in Korea].' Choi Hanki, a friend of his in Seoul, stated in one of his essays that Kim was interested in maps and regional geography from his boyhood (Choi Hanki, p. 1). From this it may be supposed that Kim had an ambition to be a geographer (and a cartographer) from early childhood.

It is believed that Kim may have moved to Seoul before the age of 20, because he completed his first major cartographic work, *Chonggudo* (*The Map of Blue Hills*), when he was only 30 (Yoon, 1991, p. 40). It would have taken considerable time to collect the data and draw such a detailed and accurate map. It would have been difficult to obtain the information included in the map from anywhere other than Seoul.

In Seoul, Kim is reported to have lived in a southern suburb, where he was befriended by Choi Hanki, the owner of a large private library and a scholar who was interested in the newly introduced western science. Kim engraved the printing block of the world map, which Choi Hanki edited and reproduced, while Choi Hanki wrote a preface to Kim's first map of Korea, *Chonggudo*. Judging from the scholarly relationships between them, it is most probable that the two were good friends, with Kim having access to Choi's library. Kim seems to have absorbed western ideas and maps through Choi Hanki.

For the printing of Kim's remarkable map, *Taedongyojido*, it seems that Choi Songhwan, a court officer, either employed him to complete the job or assisted him financially. The court officer also helped Kim produce his second edition of the regional geography of Korea, *Yodobiji*.

Kim devoted his life to producing his three maps and three different editions of his comprehensive regional geography of Korea. Those three maps are *Chonggudo* (*The Map of Blue Hills*), *Tongyodo* (*The Map of the East Nation*) and *Taedongyojido* (*The Map of the Great East Nation*). Three different editions of his Korean regional geography are *Tongyodoji* (*The Book of the Map of the East Nation*), *Yodobiji* (*The Book to Accompany the Map of the Land*) and *Taedongjiji* (*Regional Geography of the Great East Nation*). However, he was never awarded a prize for his works by the government, nor was he ever financially well off. He carved his last map, *Taedongyojido* (*The Map of the Great East Nation*) at the age of 60, only a few years before his death. It is believed that he died in Seoul in 1866, while completing his most comprehensive regional geography of Korea, *Taedongjiji*.

Owing to his fame and the scarcity of biographical data, his work and achievement have sometimes been subject to apocryphal belief. Three of the stories which circulate most frequently are given here.

It is generally believed that Kim completed his map *Taedongyojido* on the basis of his own painstaking survey of the whole of Korea. In a western cartographic journal, *Imago Mundi*, Norman Thrower and Youngil Kim have stated, 'Kim did not rely on secondary material for his information but travelled throughout Korea making his own observations' (Thrower and Kim, p. 34). However, it is now known that Chong-ho Kim did not travel Korea to survey the whole country by himself, although he may have travelled in some parts to verify his secondary map information. In fact, most of his information came from government maps on the local county, provincial or district levels. Additionally, he had access to the military maps housed in the central government through Shin Hon, a high-ranking court officer (Yi Sangtae, 1991, pp. 179–80).

It is commonly supposed that he climbed the highest peak of Korea, Mount Paektu, several times. Yet it is physically and financially unreasonable to assume that a person could travel from Seoul to Mount Paektu several times (Yi Pyongdo, p. 180). His description of the Mount Paektu district, with which he is supposed to be familiar through personal observation, is one of the least accurate areas in his maps and regional geography books. For these reasons it is suggested that he never travelled there.

He was thought to have died in prison for producing an accurate and comprehensive map which might have exposed the country to a foreign enemy. The government of the Choson Dynasty is said to have accused him of spying for a foreign country. This story, which was even known to the west, first appeared in a school textbook prepared by the Japanese colonial government for Korean students. Some now suggest that the story might have been fabricated by the Japanese colonial government to demoralize Korean pupils by convincing them that their government unwisely persecuted such a scholar. This widely circulated story is still believed by many Korean and western scholars of the present day. For instance, an article in *Imago Mundi* reports that 'Kim was later imprisoned, charged with having prepared the maps to assist the Japanese invade Korea' (Thrower and Kim, p. 34). However, as Yi Pyongdo has convincingly argued, there is no evidence to support Kim's imprisonment for his masterly work (Yi Pyongdo, p. 180). All available evidence suggests that Kim neither was imprisoned nor died in prison. As Yi Pyongdo suggests, if he had been imprisoned, the government would have confiscated his printing gear, woodblock engravings and printed maps (Yi Pyongdo, p. 180). It is almost certain that the complete set of the woodblock engravings survived the Japanese colonial government, and possibly the beginning of the Korean War (Yi Sangtae, 1991, p. 205). Even today, a block of that engraving has survived in the Sungshil University Museum in Seoul.

2. Scientific Ideas and Geographical Thought

As a cartographer and geographer, Kim made important contributions to Korean geography. His major academic achievements consisted of the three different editions of Korean regional geography and three maps of Korea.

As a regional geography of all Korea, Kim's study remained the most comprehensive and accurate into the twentieth century. With meticulous scholarship, he produced three editions of the Korean regional geography containing detailed information unavailable in other regional geography books of the time.

Kim completed twenty-two volumes of *Tongyodoji*, his first edition of regional geography, at the age of 30 in 1834. Each district treated in the book includes data concerning local history, mountains, rivers, the extent of territory, customs, population, military bases, government storehouses and a host of other topics. This edition adopted the structure of *Tonggukyojisungnam* (*The Regional Geography of Korea*), published by the government three hundred years before his time (Yi Sangtae, 1991, p. 137).

He continuously improved and revised his first book of regional geography and published twenty volumes of *Yodobiji*, his second book of regional geography, between the years 1853 and 1856. It was a somewhat abridged and rearranged version of *Tongyodoji* (Yang Pokyong, p. 166), his first book of regional geography, and was completed with financial and editorial assistance provided by Choi Songhwan (Yi Sangtae, 1991, pp. 141–2).

His third and most comprehensive work was a thirty-volume regional geography of Korea, *Taedongjiji*. He is thought to have started this work soon after the completion of his superb map, 1861, and continued working on the book until his death in 1866 (Yi Sangtae, 1991, p. 158). He borrowed from his first edition of regional geography, *Tongyodoji*, and from his second edition, *Yodobiji*, to render this final edition.

The most important contributions made by Kim may be his three maps of Korea. Until the introduction of modern topographic maps, his were adopted as the standard maps of Korea.

His first major map was the map of Korea called *Chonggudo*, which was produced in 1834. This map adopted a form of latitude and longitude to ensure that the same scale (about 1: 216,000) was applied to all districts on it. This was the most accurate and comprehensive map of Korea at the time. The topographic expression focused on the major mountain ranges and lesser mountains surrounding settlements. This topographic emphasis suggests an interest in geomancy, an east Asian art of selecting an auspicious site by evaluating landscapes. The surrounding mountain ranges (called dragons by geomancers) and the main hill in the background of a settlement are some of the most critical criteria considered by professional geomancers, for through them vital energy, which influences people, is delivered.

In the introduction to his map, *Chonggudo*, Kim argued that one does not need to draw the junction of mountain ranges on the map, since one can locate such places by observing the source of streams or rivers. This argument was an original and important cartographic contribution, although he might have derived the idea partly from the traditional geomancer's method of field observation (see below). Indeed, in a mountainous country such as Korea, Kim's theory can be safely applied.

Tongyodo was the second major map of Korea published by Kim. It remained the most comprehensive and detailed map completed in the history of Korean cartography until the introduction of modern topographic maps. It was a hand-drawn map, being the sister map of his book of regional geography, *Tongyodoji*. It is assumed that it was also intended to be a preliminary map for the printing of *Taedongyojido*, his most famous work. For this map he adopted a table of twenty-six map symbols, neatly arranged and explained, called '*Chidopyo*'. This was an innovation in Korean cartography.

Taedongyojido was the last of his three major maps and the only one to be printed (the other two were hand-drawn copies). It was printed from 126 woodblock engravings, six of which constituted the introduction to the map and close-up views of Seoul and its vicinity. Because of the difficulty of engraving on woodblocks, Kim was selective in map information. For instance, the hand-drawn copies of *Chonggudo* and *Tongyodo* had 4,712 and 5,419 entries of mountain names respectively. *Taedongyojido*, however, had only 4,572 entries of mountain names (Yi Sangtae, 1991, pp. 171–2). Views differ regarding the scale of the map, which is thought to be about 1:216,000 (Yi Wuhyong, p. 22). This map consists of twenty-two folding sections in latitudinal order from north to south. It is so accurate that the Japanese army used it in their military campaign in Korea in the Sino-Japanese War, 1894; it is still used by mountain hikers. Until the introduction of modern topographic mapping, *Taedongyojido* was the most accurate and detailed traditional map covering an entire nation printed in east Asia.

Kim is also reported to have produced *Susondo*, *Taedongyojijondo* and other maps. It is, however, not possible to verify authorship of these works.

In the geographical study of a region, Kim believed that the book and map of the area concerned are an inseparable pair. That is why he completed the three maps of Korea and three matching editions of its regional geography. He felt that without one, the other could not fulfil its intended aim.

The introduction to each of his three maps expresses the idea as follows: in *Chonggudo Pomrye* (*Notes on Chonggudo*), he stated, 'a book of regional geography is the base for a map of the area concerned'. In *Tongyodojiso*, an introduction to the regional geography to accompany the *Tongyodo Map*, he made a similar statement:

> Generally, a book of regional geography should be accompanied by a map of the area concerned. . . . With a map one observes the landscape [landforms] of the world and with a book of regional geography the [social] institutions and [people's] material achievements through time are traced. Therefore, these two are of great importance in governing the nation.

In *Chidoryusol*, the third and final map in the introduction to *Taedongyojido*, he stated:

> Someone said that . . . generally with a map the configuration of a land is observed and with a book of regional geography its quantity [of the configurations] is measured. To the left of a scholar a map was placed and to the right, a book [of regional geography] . . .

From these statements we recognize that, in his view, the book was the source for a map. He also thought that a map had an advantage over a book in its visual expression of the configuration of a land.

He successfully derived scientific concepts from geomancy, which contains elements of folk-lore, superstition and religion (Yoon, 1976, p. 234). It is assumed that Kim was exposed to geomantic ideas then prevalent in Korea. This is reflected in his emphasis of geomantically important land forms in, for example, his map *Taedongyojido* (Yoon, 1991, pp. 40–6). In *Chonggudo Pomrye*, he suggests that the mountain ranges and watercourses were the flesh (muscle and bone) and blood vessels of land. This was a view typically held by geomancers. The successful and ingenious application of the geomantic idea in his work is evident in the expression of mountain ranges in *Taedongyojido*, which was an improvement on the geomantic map.

Probably by learning from geomantic maps, Kim developed a sophisticated skill in the expression of land forms on maps not seen previously. Traditionally, in East Asia, land-form expression was by use of realistic drawing (for example, for mountains, shapes of mountain peaks were drawn). Geomantic maps adopted well-developed methods of land-form expression, distinguishing mountain peaks, slopes and foothills (Yoon, 1976, p. xxiii, and 1992a, p. 14). There is a remarkable resemblance between geomantic maps and Kim's *Taedongyojido* in topographic expression. In *Taedongyojido*, however, unlike the geomantic map, mountain ranges were expressed with thick lines, the thickness of the line indicating the size of the range.

The second of his cartographic contributions was an adaptation and application of the geomantic idea that 'between the two watercourses there is a mountain range [upland] and between the two mountain ranges there lies a watercourse [valleys]' (Hsu Shan-chi and Hsu Shan-shu, Vol. 1, Section 1). This statement, a well-known geomantic proverb, became a guiding principle for geomancers in their observations of landscapes. By applying it, Kim surmised that, as long as the river pattern is accurately drawn, it is unnecessary to express every junction of the mountain range, because the source of a stream or river has its origin there.

In this way, Kim made timeless contributions to Korean geography. He is admired by geographers and others alike in Korea as the most important cartographer and geographer of the country.

3. Influence and Spread of Ideas

Kim's *Taedongjiji* is not only the most comprehensive and accurate regional geography of Korea, but it also includes information on the time not available in other sources. His work is used even today as a reference book for geographers, historians, economists, sociologists and other scholars who study the Korean Choson Dynasty. *Taedongjiji* is now published as a facsimile edition because the book is in great demand. Kim's maps, especially *Taedongyojido*, have been reprinted frequently, and as recently as 1991. Kim's works represent the pinnacle of achievement of traditional Korean geography.

Contemporary Korean geographical ideas and approaches represent an importation of the western tradition rather than the continuation of traditional Korean geography. Kim's scientific ideas, cartographical skills and geographical approaches are hardly utilized by Korean geographers today. Therefore Kim's influence on present-day Korean geographical scholarship is not as significant as it might otherwise have been. Today, there are some Korean geographers who have contemplated the revival and incorporation of Kim's achievements and approaches into modern geographical scholarship: this is an important challenge.

Because of his scholarly achievements, Kim became a man admired by generations of Koreans. He is the only geographer to be honoured as a cultural personality of the month (April 1991). A major street in the city of Seoul has been named after him. Kim's life and work is included in the South Korean school textbook for fifth-grade primary school. Indeed, all Koreans with a primary-school education have heard of him. In most Koreans' minds he is remembered as a geographer who made an outstanding contribution to Korean culture.

Bibliography and Sources

1. REFERENCES ON CHONG-HO KIM

Choi Hanki, *Chonggudoje* [*Introduction to Chonggudo, Map of Korea*].

Hsu Shan-chi and Hsu Shan-shu (Ming Dynasty), *Jen-tzu hsu-chih*, Hsin-chu, Chu-lin Shu-chu, 1969.

Thrower, N.J.W. and Kim, Y., 'Dong-Kook-Yu-Ji-Do: a recently discovered manuscript of a map of Korea', *Imago Mundi*, Vol. 21 (1967), 30–49.

Yang Pokyong, 'Kosanja Chijiui Hyondaejok Pyongka' ['An evaluation of Kim Chong-ho's regional geography from a modern perspective'], *Chirihak*, Vol. 26, No. 2 (1991), 164–73.

Yi Pyongdo (ed.), 'Kim Chong-ho, Chonggudo, Taedongjiji, Taedongyojido', in *Hangugukui Kojonpaekson* [*100 Selections of Korean Classics*], Seoul, The Tong-a Ilbo, 1969, 179–84.

Yi Sangtae, 'Kosanja Kim Chong-howa Tongyodo' ['Kim Chong-ho and the Tongyodo Map'], in Yi Wuhyong (ed.), *Taedongyojidoui Tokdo* [*Reading the Taedongyojido Map*], Seoul, Kwangwudang, 1990, 3–7.

Yi Sangtae, 'Chosonsidae Chidoyongu' ['A Study of the Maps during the Choson Dynasty'], Ph.D. thesis, Tongguk University, 1991.

Yi Wuhyong, *Taedongyojidoui Tokdo* [*Reading the Taedongyojido Map*], Seoul, Kwangwudang, 1990.

Yoon, H., *Geomantic Relationships Between Culture and Nature in Korea*, Taipei, Orient Culture Service, 1976.

Yoon, H., 'Taedongyojidoui Chido chokboronjokin yongu' ['A carto-genealogical study of *Taedongyojido*'] [in Korean with English abstracts], *Yoksa Munhwa Chiri*, Vol. 3 (1991), 37–47.

Yoon, H., 'Taedongyojichondo Somune taehan Yebikochal' ['A preliminary study of "The Preface" to the Taedongyojichondo Map'], *Yoksa Munhwa Chiri*, Vol. 4 (1992), 97–107.

Yoon, H., 'The expression of landforms in Chinese geomantic maps'. *Cartographic Journal*, Vol. 28, No. 1 (1992a), 12–15.

Yoon, H., 'The traditional standard Korean maps and geomancy', *New Zealand Map Society Journal*, No. 6 (1992b), 3–9.

Dr Hong-key Yoon is a Senior Lecturer in Geography at the University of Auckland, Auckland, New Zealand.

Chronology

c. 1804	Born in Hwanghae Province in Korea (presently, North Korea)
1834	Completed *Tongyodoji*, his first book of regional geography of Korea; completed *Chonggudo*, the first major map of the whole of Korea

c. 1851–56 Completed *Yodobiji*, his second book of regional geography of Korea

c. 1856–61 Finished drawing the most comprehensive map of Korea, *Tongyodo* (this map became the base map of *Taedongyojido*)

1861   Printed the first edition of *Taedongyojido* from woodblock engravings. It was the most comprehensive map of Korea ever printed then

1864   Printed the second edition of *Taedongyojido* from woodblock engravings

1866   Died in Seoul while completing his most comprehensive book of regional geography of Korea, *Taedongjiji* (completed all except Pyongan Province), based on his previous works

Paul Meuriot

1861–1919

Gilles Montigny

In a period marked by the emergence of the French School of Geography, typified by Raoul Blanchard, Albert Demangeon, Lucien Gallois, Emmanuel de Martonne, Jean Brunhes and others, and dominated by the influence of Vidal de la Blache, Paul Meuriot was generally seen as a lesser figure. However, even though he belonged to the group studying social statistics, his contribution to the field of geography should not be underestimated. He considered facts whose significance was relatively undervalued by the Vidalians: the causes and consequences of urbanization; the spread of urban centres; the intensity of urban growth; and the specific characteristics of the modern city as compared to the old. In the period between 1890 and 1920, Meuriot – along with another geographer who did not belong to the School, Elisée Reclus – without question took observation and reflection on urban phenomena further than all others.

1. Education, Life and Work

Paul Marie Gustave Meuriot was born in Bar-sur-Aube on 18 December 1861. After completing his secondary education in several provincial high schools, he went to Paris, where he prepared for the admissions examination of the Ecole Normale Supérieure at the Lycée Louis-le-Grand. He failed the examination twice, in 1881 and in 1884, but in 1889 obtained his *agrégation* in history. While a teacher at high schools in Belfort, Nevers and Amiens, he undertook studies of contemporary migrations. In 1893, he received a prize awarded by the Academy of Moral and Political Sciences for this work, which marked the beginning of a larger project that would culminate five years later in a doctoral thesis on urban centres in contemporary Europe. The work remains one of the most significant texts in French geography on the subject.

Having gained the attention of E. Levasseur somewhat earlier, Meuriot became a member of the Statistics Society of Paris in December 1896. The poor reception which met his Latin thesis, with historical geography, prevented him from obtaining a university position, however, and he carried out the rest of his career at the Lakanal High School in Sceaux, where he was appointed in 1897. He none the less continued for the rest of his life to engage in research of broad range and the highest quality. Between 1897 and 1919, he published approximately 160 articles, a great many of which dealt with urban and demographic topics.

Thanks to Meuriot's fine command of foreign languages, these studies deal not only with France, but also with other countries and continents. Some of his subjects were quite original, especially for his day. These are devoted to electoral geography and the effects of systems of parliamentary representation; the distribution of languages, nationalities and religions; the economic development of neighbouring countries (Germany, England and Luxembourg); illiteracy and the availability of schooling; diverse economic questions (income tax in Great Britain, national budgets, the evolution of agricultural prizes, inheritance taxes); and the democratization of speed (using the example of the railway). He regularly reviewed foreign publications in statistics and the progress in the organization of statistical services in many countries. His scholarly work also includes the publication of the results of the census undertaken in the year II, mandated by the Convention, which had never before been done (Meuriot, 1918a).

These articles, numbering 162 in all, may be classified thematically as follows: the study of urbanization and urban or statistical centres relating to cities (28); results of national statistics and inquiries relating to foreign statistical services (28); economic questions (23); political and electoral geography (17); the distribution of languages, nationalities and religions (17); studies involving general demographic phenomena (15); reviews (13, of which 5 are related to urban questions); geography of education and instruction (8); other (13).

The bulk of these articles were published in the *Journal de la Societé de Statistique de Paris* (*Journal of the Statistics Society of Paris*). A few, particularly those dealing with the measurement of urban centres and the comparison between the various national urban statistics, appeared in the *Bulletin de l'Institut International de Statistique* (*Bulletin of the International Institute of Statistics*); Meuriot belonged to the institute from 1909 on. Recognised as one of the best statisticians of his day, he assumed the presidency of the Statistics Society of Paris in January 1919. This was also the year in which he published his last two articles, including the groundbreaking essay on the concept of the city, in *La Vie urbaine* (*Urban Life*), the journal of the new Parisian Institute of Urban History, Geography and Economy. When Paul Meuriot died in Paris on 3 December 1919, his election to the Academy of Moral and Political Sciences was imminent.

2. Scientific Ideas and Geographical Thought

Meuriot's entire scientific production is based on a conception of geography initially developed by his mentor, E. Levasseur. This was an experimental geography, relying upon numerous and detailed statistical observations. For these geographers, statistics were the surest means of accounting for human diversity and for going beyond irregularities tied to individual freedom. With the help of figures, it became possible to describe as precisely and as faithfully as possible the realities of the world and of its evolution. Distinguishing itself from German geography, especially that of C. Ritter, the statistical approach refused to consider natural environment as playing the principal role, which it attributed, instead, to people. The task of the geographer thus consisted in describing and explaining settlements, population centres, the wealth of states and of their regions as a function of populations, natural resources and the economy. In so far as the data gathered in this way would allow, it was possible to formulate geographic laws: Levasseur described several such laws involving the development of urban centres, and Meuriot furthered this research. Geography thus became complementary, significantly propped by statistics, relating to history and political economics. All of these characteristics are masterfully brought together in Meuriot's work.

Towards 1903, a few years after coining the term 'urban geography', J. Brunhes like other Vidalians adopted the term 'human geography'. Meuriot devoted his life's work to this fledgling branch of geography that was only just beginning to distinguish itself and gain its autonomy; he carried out investigations in areas thus far only minimally explored. The best evidence of this is provided by the content analysis of his articles in the preceding section. This constitutes the second principal characteristic of the author's approach, based on a very broad conception of what the field of human geography ought to be. His was a broad and at the same time innovative approach; again and again Meuriot cut the figure of the pioneer. As the precursor of political and electoral geography (the work recognized as such being the *Tableau politique de la France de l'Ouest* [*Political Chart of Western France*] written by André Siegfried in 1913), he was also that of the geography of religion at the time when the sociology of religion was just beginning (in the work of M. Weber, E. Troeltsch and E. Durkheim), and when the first attempts at a systematic cartography of religion were still far off (the first article by G. Le Bras on this question dates from 1931). He was also among those who, after E. Levasseur and at the same time as D. Pasquet, A. Vacher and R. Blanchard, founded urban geography.

Throughout his publications and his activity within the International Statistics Institute, Meuriot displayed clear information regarding cities and urban centres, whether French, European or world. Historians of urbanization in the nineteenth and early twentieth centuries increasingly make use of his work, which constitutes a first-rate primary source. Recognition was late in coming, but well deserved. He took up and discussed the respective roles of the principal factors of urbanization (the relative overpopulation of rural areas, regimes of land ownership, relative salaries between agriculture and industry or the decline of rural industries). He also considered the material consequences (the development of suburbs, for example) and the social consequences (in their economic, religious, political and demographic dimensions) of this phenomenon.

Beginning in 1897, he more than anyone attempted to develop the concept of the city. Refusing to adhere to morphological criteria, legal criteria, or criteria relating to the division of work between cities and rural areas, and aware of the difficulty involved in separating the city centre from other communes which constituted with it the urban centre, he proposed a definition which encompassed both. This is contained in the following formula: 'The modern urban centre is thus nothing other than a dynamism maintained by the indefinite growth of its population and the equally indefinite extension of its region' (Meuriot, 1919c). In several sessions of the International Statistics Institute, Meuriot made proposals in an attempt to resolve a practical problem: how could the great urban centres be demarcated in a way that would make possible comparative studies on an international scale? He proposed measuring the force of attraction of the city centre in order to define a perimeter (1909 session), observing the origin – on the basis of immigration – of the population of suburbs (same session), and introducing a criterion of density balanced by the distance from the city centre (1911 session), to take into account the growth of the population in a neighbouring zone (1913 session). If it is difficult to determine a criterion for the city centre, the suburb phenomenon is better understood. The suburb is not 'any longer the region neighbouring the great city, but the one whose population is no more than the prolongation of the city'. With E. Reclus and D. Pasquet, Meuriot was one of the few French geographers of his day to demonstrate the magnitude not only of the city centre, but of the phenomenon of urban concentration.

In the mapping out of the urban centre, the periphery poses a particular problem. Anticipating a dichotomy which has become common in our own day, Meuriot also attempted to define the centre of a city or urban centre. His method involves taking note of the distribution of the resident population and localizing, as for a centre of gravity, the 'mathematical centre' of the population. The method may be questionable, but it is able to account for a possible displacement of the centre, and it is a good example of the way in which statistics can contribute, if abstractly, to an objective description of urban realities.

His use of statistics led Meuriot to accept an idea held by Levasseur according to which urban development is subject to general laws. He himself added to the list of these laws, describing the development of foreign immigration and that of urban populations as a relationship of necessity. He argued that the same relationship held in the case of altitude and urban settlements. This did not, however, prevent him from refuting the most famous of these laws: the 'law of westward development of great urban centres'. According to Meuriot, this law is disproved by observation, and particularly by the determination of the mathematical centre, which we have discussed above.

3. Influence and Spread of Ideas

As mentioned in the introduction, Meuriot, like E. Reclus, was marginalised by the Vidalians. His work was hardly consulted by them. Consequently, it was not widely disseminated beyond the (albeit international) circle of social statisticians. As his teaching was limited to high schools, Meuriot did not have any direct disciples, although he himself had been one (of Levasseur).

It would be false, however, to claim that his work has had no influence. In the 1910s, the question of the international comparison of urban populations and of the size of urban centres was far from being already settled. It was taken up again in the 1930s with reference to Meuriot's work, once again as part of the research conducted in the International Statistics Institute. The director at the time of the Census Bureau of General Statistics of France, Henri Bunle, actively pursued the subject. Specialists at the National Institute of Statistics and Economic Research (INSEE) took it up again when multi-communal urban centres were circumscribed with precision for the censuses of 1954 and 1962 (these definitions are still used). At the United Nations, specialists are still working toward a definition of urban concentration that may be applied world-wide.

In a general way, Meuriot broke the ground in quantitative studies of the urban phenomenon and, to his credit, he engaged in such research at a time when French geographers who were interested in urban geography, following the lead of R. Blanchard, produced monographs of cities with limited or no recourse to statistics.

Bibliography and Sources

1. REFERENCES AND OBITUARIES ON PAUL MEURIOT

Barriol, A., 'Nécrologie', *Journal de la Société de Statistique de Paris*, Vol. LXI (1920), 14–16.

Berdoulay, V., *La formation de l'école française de géographie 1870–1914*, Comité des travaux historiques, Paris, Bibl. Nationale, 1981, p. 166.

Gaudin, J.-P., *Desseins de villes. Art urbain et urbanisme*, Paris, L'Harmattan, 1991, 19–20, 23–5, 168.

Huber, M., 'Nécrologie', *Bulletin de l'Institut International de Statistique*, Vol. XXI (1924), 368–9.

Montigny, G., *De la ville à l'urbanisation. Essai sur la genèse des études urbaines françaises*, Paris, L'Harmattan, 1992, 100–3, 323–4; bibliographie complète de P. Meuriot, 363–72 (164 titles).

Roncayolo, M. and T. Paquot, *Villes et civilisation urbaine, XVIIIe–XXe siècle*, Paris, Larousse, 1992, 21–34, 440–4.

Archives nationales: 61 AJ 169 (dossier P. Meuriot).

2. BIBLIOGRAPHY OF PAUL MEURIOT'S MOST REPRESENTATIVE WRITINGS

1897   *Des agglomérations urbaines dans l'Europe contemporaine*, Paris, Belin, 475 pp.

1897   *Qualem Britaniae formam veteres geographi sibi fuxerint* (Latin thesis), Paris, Belin, 63 pp. and maps.

1898   'Les étrangers dans l'Empire allemand, d'après le dernier dénombrement', *J.*, Vol. 39, No. 9, 301–9.

1899   'La carte électorale de l'Empire allemand', *J.*, Vol. 40, No. 6, 201–9.

1900   'Un exemple de dépopulation rurale. Le département de la Somme', *J.*, Vol. 41, No. 8, 265–71.

1901a   'La démographie d'une petite ville au XIXe siècle (Bar-sur-Aube)', *J.*, Vol. 42, No. 12, 407–12.

1901b   'De l'expression Diaphragma dans l'histoire de la géographie ancienne', *Congrès d'histoire des sciences de Paris de 1900.*

1902   'Du centre mathématique d'une population', *J.*, Vol. 43, No. 8, 268–71.

1904a   'Les divers modes d'évaluation de la population de Londres', *J.*, Vol. 45, No. 4, 144–50.

1904b   'La répartition de la population en Suisse par altitude', *J.*, Vol. 45, Nos. 5 and 6, 180–2, 202–11.

1904c   'De l'influence des migrations internes sur la répartition des langues et des cultes en Suisse', *J.*, Vol. 45, No. 8, 274–82.

1904d   'La transformation des agglomérations urbaines sous l'influence des facteurs physiques, économiques, administratifs et sociaux', *Mémoire Acad. des sc.*

1905   'La répartition des langues en Belgique', *J.*, Vol. 46, No. 10, 238–45.

1908a   'La métropole de l'Amérique latine. Buenos Aires au début du XXe siècle', *J.*, Vol. 49, Nos 2 and 3, 56–67, 86–92.

1908b   'La petite ville française', *J.*, Vol. 49, Nos 7 and 8, 235–40, 245–53.

1909a   'La démocratisation de la vitesse', Part 1, *J.*, Vol. 50, No. 1, 14–20.

1909b   'Les migrations internes dans quelques grandes villes', *J.*, Vol. 50, No. 8, 390–400 (maps).

1909c   'De la mesure des agglomérations urbaines', *Bulletin de l'Institut International de Statistique*, Vol. 18, Nos 1 and 2, 82–94, 133–7.

1910a   'L'annuité successorale en France et la population (1876–1906)', *J.*, Vol. 51, No. 1, 11–22.

1910b   'Les petites communes françaises', *J.*, Vol. 51, No. 3, 91–7.

1910c   'Les villes prussiennes' (following the work of H. Silbergleit), *J.*, Vol. 51, No. 7, 277–8.

1911a   'Rapport sur la question de la mesure des agglomérations urbaines', *Bulletin de l'Institut International de Statistique*, Vol. 19, No. 1, 157–61.

1911b   'Les élections anglaises de 1910', *J.*, Vol. 52, No. 5, 184–94.

1912a   'Dans quel sens se développent nos grandes villes européennes?', *J.*, Vol. 53, No. 5, 238–47.

1912b   'La démocratisation de la vitesse', Part 2, *J.*, Vol. 53, No. 6, 280–9.

1912c   'La population étrangère en France (1851–1911)', *J.*, Vol. 53, No. 12, 555–65.

1913a   'Quinze ans d'Income-Tax', *J.*, Vol. 54, No. 4, 187–202.

1913b   'Rapports sur la mesure des agglomérations urbaines, le mouvement migratoire à Paris et à Londres à l'époque contemporaine, l'état présent de la statistique religieuse en Europe et hors d'Europe', *J.*, Vol. 54, No. 11, 590–2.

1914a   'La population de jour de la Cité de Londres', *J.*, Vol. 55, No. 2, 96–7.

1914b   'Du criterium adopté pour définir la population urbaine', *J.*, Vol. 55, No. 10, 418–30.

1914c   'L'analyse théorique sur la ville et la réflexion des économistes du XVIIIe siècle', *Séances et travaux de l'Acad. des sc. mor. et pol.*, May 1914.

1915a   'De la valeur du terme de banlieue dans certaines métropoles: Paris, Berlin, Londres', *Bulletin de l'Institut International de Statistique*, Vol. 20, No. 2, 320–9.

1915b   'Le mouvement migratoire contemporain à Paris et à Londres', *Bulletin de l'Institut International de Statistique*, Vol. 20, 306–19.

1916   'Du calcul de la population par feux', *J.*, Vol. 57, No. 11, 455–7.

1917a   'La langue celtique dans les îles britanniques', *J.*, Vol. 58, No. 6, 171–84.

1917b   'La répartition des langues en Belgique', *J.*, Vol. 58, No. 11, 339–53.

1917c   'Le progrès de la Corée sous la domination japonaise', *J.*, Vol. 58, No. 11, 353–8.

1918a   'Le recensement de l'an II', *J.*, Vol. 59, Nos 2 and 3, 34–56, 79–99.

1918b   'Le suicide à Paris avant et pendant la guerre', *J.*, Vol. 59, No. 10, 269–85.

1919a   'Le baccalauréat. Son évolution historique et statistique des origines (1808) à nos jours', *J.*, Vol. 60, Nos 1 and 2, 8–35, 67–89.

1919b   'Du concept de ville autrefois et aujourd'hui', *La Vie urbaine*, Vols 1 and 2, 145–54.

1919c   'La densité de la population de Paris: densité mathématique et densité réelle', *La Vie urbaine*, Vol. 4, 457–68.

Gilles Montigny is an agrégé *of the university and a doctor of social sciences from the Ecole des Hautes Etudes en Sciences Sociales.*

Chronology

1861	Born in Bar-sur-Aube, 18 December
1884	Earned the *agrégation* in history
1884–88	Teacher ('professor') at Belfort High School
1888–90	Teacher at Nevers High School
1890–97	Teacher at Amiens High School
1893	Awarded the Bordin Prize of the Academy of Moral and Political Sciences
1896	Admission to the Statistics Society of Paris
1897–1913	Teacher at Lakanal High School (in Sceaux)

1898	Awarded a doctoral degree upon completion of a dissertation on urban centres in contemporary Europe (first published work)
1904	Awarded the Budget Prize of the Academy of Moral and Political Sciences
1909	First participated in the research of the International Statistics Institute
1910	Became a member of the Advisory Council of the Statistics Society of Paris
1915	Substitute teacher at the Carnot High School in Paris.
1919	President of the Statistics Society of Paris; died in Paris, 3 December

Translated by L. Scott Lerner

John Linton Myres

1869–1954

William A. Koelsch

Portrait of J.L. Myres reproduced by
permission of the British Academy.

John L. Myres may be described (by analogy with the well-known characterization of the geologist N.S. Shaler) as 'a classicist by training and yet a geographer in instinct'. From his student days at Oxford, where he was one of Mackinder's earliest hearers, until the publication of his collected essays in classical geography a year before his death, geographical perspectives informed Myres' thinking, scholarship, teaching and institutional activity. He was also a key figure in the task of raising the status of geography in British schools, universities and scientific institutions in the early twentieth century.

1. Education, Life and Work

John L. Myres was born in Preston, Lancashire, on 3 July 1869, the only son of the Rev. William M. Myres, Vicar of St Paul's, Preston, and his first wife Jane (Linton), herself the daughter of a clergyman. His grandfather and great-uncle were both mayors of Preston, and Myres retained a lifelong interest in the north country of his rootage. He was related both to Percy Maude Roxby, who began his career as a professional geographer in 1904 at the University of Liverpool, and to the husband of novelist Rebecca West.

Myres had been fascinated with archaeology as a child, initially through looking at the pictures in his father's copies of the reports of the Palestine Exploration Fund. His formal schooling began in 1879 at the Preston Grammar School, but with his father's removal to the rural living of Swanborne, Buckinghamshire, he entered boarding school at Thorpe Mandeville. In 1882 he won a scholarship to Winchester, where the curriculum was expanding to include the sciences, so that the young Myres gained the foundation in chemistry, physics and geology as well as classics which he was to offer for admission to Oxford in 1888. While at Winchester, Myres was in charge of the school's museum, cataloguing its fossils and caring for its antiquities from Cyprus, an area in which he was to earn his spurs as a classical archaeologist. Myres read widely in Winchester's library, and found there further stimulus toward his eventual career in his encounter with Heinrich Schliemann's account of his discoveries at Troy.

Myres entered New College, Oxford, in 1888, the leader in scholarship among the Winchester entrants of that year. While an undergraduate he heard the lectures of the recently appointed Halford Mackinder in geography, challenged the Oxford anthropologist Edward B. Tylor to develop a course of lectures on the relation of anthropology to ancient history, listened with fascination as Gladstone lectured on Homer before the Oxford Union, and kept up his interests both in science and in history, becoming president of the Oxford Architectural and Historical Society.

His first experience of a real dig came at the site of a Roman market-town at Alchester, Oxfordshire. After graduating, Myres earned a fellowship at Magdalen College and scholarships in both geology and classics. He had made a brief trip to Greece in 1891 and was fascinated by the Greek countryside and its rural inhabitants. In 1892 he headed for the Greek islands and Caria, explored Crete with Arthur Evans, made the first important correlations between Cretan and Egyptian civilizations on the basis of pottery comparison, and conducted his first excavations on Cyprus – all before the age of 25.

During his two years in Greece in this context Myres developed a knowledge base embracing both land and people, 'and from then on,' wrote one of his former students, 'the Aegean was as vividly alive for him in classical writers as it was while he was actually on its hills and seas'. In 1896 Myres was elected a student of Christ Church College and University Lecturer in Classical Archaeology at Oxford; part of his responsibilities was to lecture on the historical geography of the eastern Mediterranean.

Myres left Oxford in 1907 to become the first Gladstone Professor of Greek and Lecturer in Ancient Geography at the then relatively new University of Liverpool, to which he was attracted in part because of its opportunities for educational innovation not then abundant at Oxford. Because the damp climate of the seaport affected his wife's health, however, he moved back to Oxford in 1910 as the first Wykeham Professor of Ancient History. The chair, which carried with it particular responsibilities for lecturing on ancient Greece, was attached to New College, with which Myres was to be associated (part of the time as librarian) for the rest of his life, even after his formal retirement from the chair in 1939. His inaugural lecture, entitled 'Greek lands and the Greek people', accurately reflects the dual interests that informed his scholarly career.

In 1914 Myres became the first Sather Professor of Classical Literature at the University of California, Berkeley, refusing an invitation to stay on there as Professor of Ancient History. From 1915 to 1918 he was occupied with war work, first as a special intelligence officer for the Admiralty and later as an information and passport control officer in Athens. In between these assignments he served as commander of the tug *Syra*, conducting raiding parties on the Turkish coast, which brought Myres detailed knowledge of the Dodecanese and the landscapes of ancient Asia Minor.

Back at Oxford he became active in university politics, including the effort, begun before the war, to establish an honour school of geography. He was elected a general secretary of the British Association for the Advancement of

Science in 1919, lectured in America both in 1926, at Wesleyan University, and in 1927, as Sather Professor for the second time; both of these series of lectures resulted in significant books. He was active in the affairs of a number of learned societies, including serving as president of the Royal Anthropological Institute, 1928–31, and was a founder or one of the secretaries for a number of other anthropological organizations. He was chairman of the board of the British School at Athens between 1934 and 1937. During his lifetime he received honorary degrees from the Universities of Wales, Manchester and Witwatersrand and, on the occasion of its centennial, Athens.

Myres' formal retirement from the Wykeham chair coincided with the outbreak of the Second World War. Despite his age and gradual infirmity owing to arthritis and failing eyesight, he went to work editing handbooks of naval intelligence and writing a handbook on the Dodecanese. He edited the *Annual* of the British School at Athens and continued to edit *Man*, which he had founded in 1901, to make up for the absence of younger men on wartime service. As librarian of New College he supervised the building of a new library building and reorganized its contents. After the war, Myres published three major books, two of them in his eighty-fourth year. A fourth, on which he was working during his last years, was posthumously published. He died in Oxford on 6 March 1954, leaving his wife of nearly sixty years (Sophia, d. 1960); they had two sons, one of whom predeceased them, and one daughter. The surviving son, J.N.L. Myres, was Bodley's Librarian at Oxford from 1948 to 1965 and a noted scholar of early Britain.

2. Scientific Ideas and Geographical Thought

In his late years Myres recalled that, in the examination for his admission to New College, he had been asked 'In classics, was it the language or the peoples that interested me'; his reply was 'the peoples', and he commented that 'I have never repented of this answer.' To this, on his first trip to Greece, he added 'the lands', for as he also recorded late in life, the trip had 'changed my whole outlook on ancient history, and especially on Greece and Greek lands. Here was a real country . . . inhabited still by a people who retained much of their ancient habit and outlook.' His whole scholarly life was spent in studying Greek lands and Greek peoples in concert, contrary to the increasing tendency in the early twentieth century to give pride of place to philology among the classical disciplines. In his inaugural lecture for the Wykeham professorship, Myres deplored the narrow literary study of the ancient world (which, at Oxford, did not begin to give way until the 1930s) and championed the application of the newer field sciences (geography, archaeology, anthropology, geology) in the study of antiquity.

Myres published his first important geographical paper, a reconstruction of the mental maps of Herodotus, in the *Geographical Journal* in 1896; parts of this essay were revisited in the third chapter of his penultimate book, *Herodotus: Father of History*, in 1953. Ten years earlier, Myres had published an essay entitled 'The ancient shape of Attica' using the same approach. In each of these, Myres is drawing on a nineteenth-century mentalistic tradition familiar to classicists and historians but somewhat different from the materialistic emphases of most° geographers of his time and since. Indeed, Myres insisted throughout his geographical and other writings on the primacy of mind, even claiming that geography did not exist until one left physiography and natural history and began to examine the interaction of mind with its external environment. In that approach, as well as insisting (along with Mackinder) that geography had a place among the humanities, Myres, like John K. Wright, may be seen as a figure in the antecedents of cognitive humanistic geography.

Myres published a number of articles in classical geography in both geographical and other periodicals in his lifetime, as well as comments, sections of books, and geographical items for encyclopaedias, including the classic eleventh edition of the *Encyclopaedia Britannica*. Among his brief geographical contributions was the article on Tripoli for Hugh Robert Mill's *International Geography*, first published in 1899. But almost all of Myres' writings have some geographical component, including his Wesleyan University lectures later published as *The Political Ideas of the Greeks*, his contributions to the *Cambridge Ancient History*, and his numerous writings on anthropology.

During the 1920s and 1930s Myres often attended and sometimes gave papers at meetings of British geographical organizations: the Geographical Association, the Royal Geographical Society, the Royal Scottish Geographical Society, and the geography section of the British Association for the Advancement of Science. A revised collection of almost all the major geographical papers, *Geographical History in Greek Lands*, was published to mark his eighty-fourth birthday in 1953. Myres' scholarly work has been largely forgotten by geographers, but as recently as 1993 the noted British classical scholar Oswyn Murray described sections of *Geographical History* as still the best introduction to ancient Greek social geography. Murray also commended Myres' Admiralty handbooks for the region, and characterizes his late essay 'Persia, Greece and Israel' as 'the fundamental discussion of the relations between Persia and her subject peoples'.

Myres made no lasting original contributions to geographic thought, yet his interactions with geographers served to model certain aspects of historical and humanistic geography which maintained the distinctive regional and historical approaches of British human geography. In formulating his position regarding geography he was influenced most clearly, among British geographers, by Mackinder, his teacher and friend; by Andrew John Herbertson, who established the regional emphasis of Oxford geography; and by H.J. Fleure, who was closely associated with Myres both in the Geographical Association and in his interests in prehistory and anthropology. But Myres also owed an important intellectual debt to his readings of the classical geographers and historians. In some of his work, especially in his Herbertson lecture of 1935, he owes something to Patrick Geddes.

Through his regular interaction with geographers Myres moved with them from a preoccupation with race and environment (a position which informed both his early work, *The Dawn of History*, and his major treatise, *Who Were the Greeks?*) to regional geography *à la* Herbertson (especially the late Herbertson who wrote of consciousness, regional psychology and the 'mental and spiritual environment') to geography as a study of distributions. He advocated the field study of landscapes, most notably in a charming essay (enlarged from a radio broadcast) entitled 'Wayside geography', and the importance of style in the presentation of geographic research.

Myres also aided the formulation of new definitions and models of historical geography as the study of regional environments of past times, not only by his example but through an important conference on the subject over which he presided at the London School of Economics in 1932. Although at times he made definitional distinctions between historical geography and geographical history, his major aim was to promote and to exemplify convergent research from both sides on the changing physical and human environments of historical events and processes. His efforts along these lines were so successful that he was accepted by geographers as a geographer; after his death, the head of the School of Geography at Oxford wrote of him that 'geography has lost a scholar of the first rank'.

3. Influence and Spread of Ideas

In his sketch of Myres in the *Dictionary of National Biography*, John Boardman remarks that 'He founded no school', but that he had shown 'how historian, archaeologist, anthropologist, and geographer should combine their skills in the study of antiquity'. Boardman also observes that Myres was 'generous and kindly' in his interactions with younger scholars, and that 'his work must be judged not only by what he wrote but by what he inspired in others, by example or casual precept'.

Boardman's remarks are borne out by the testimony of numerous others, whether or not Oxford students. Myres' lectures were heard by few, partly because he had a habit of speaking into his beard, and partly because they were viewed by many as eccentric to the specialized literary focus of classical education in the period when he was most active as a university lecturer. Yet for those who sought him out, he 'looked, and was, legendary'. As one student observer recounts,

> He would preface his first lecture with 'I hope none of you are going to waste your time taking notes!'; and so we sat back and listened – the eight or ten of us who had the good sense to attend lectures so little directed towards Schools – and would be astonished at the overflowing of a mind of such exceptional range and curiosity.

During Herbertson's time at Oxford, Myres lectured part-time in the School of Geography on ancient geography, but his influence on individual geography students of the older generation was probably as much through his writings and presence at geographical meetings as directly through his university lectures at Oxford or Liverpool. Certainly the older generation of British geographers learned from him. H.C. Darby has pointed out that in 1928 Myres was elected President of Section E (Geography) of the British Association for the Advancement of Science, then the principal research-oriented geographers' forum, and has said of Myres that his 'interests were so geographical that we almost forgot he was a professor of ancient history'. The late T.W. Freeman reminds us that Myres was 'a very firm supporter of the Geographical Association and of geography in general'. R.W. Steel writes that Myres 'campaigned resolutely . . . for the establishment of a [geography] chair' at Oxford, and that his 1911 book *The Dawn of History* 'was an outstanding example of geographical work done by an ancient historian'. (Indeed, E.C. Semple used it as reserve reading for her 1920s Mediterranean course at Clark.)

Myres had encouraged geography's expansion under Roxby while he was still at Liverpool, and from his new Oxford base after 1910 he continued to push both for the later endowed chair, the John Rankin Professorship of Geography, to which Roxby was appointed in 1917, and for the establishment of the Honour School at Liverpool, the first in any British university. For some years he served as an external examiner in geography at Liverpool.

It was at Oxford itself that Myres exercised the most direct personal influence. Myres had been close to Herbertson, and lectured in the Summer Schools which Herbertson conducted for teachers before the First World War. Myres and some other classicists and historians had pushed for the establishment of a professorship in geography and an associated honour school, though the outbreak of war aborted those efforts. In the 1920s Myres resumed this track, working to organize support from modern historians and anthropologists. When in 1931 his efforts achieved success, Myres became first chair of the Board of Examiners of the new School of Geography.

Myres had also become Honorary Secretary of the British Anthropological Association, and had used his position to persuade Oxford to establish an Honour School of Anthropology. When an endowment was found for a Professorship of Geography at Aberystwyth in the same year as that at Liverpool and the geographer-anthropologist H.J. Fleure was appointed to it, Myres and Fleure, who was also Honorary Secretary of the Geographical Association, began a long and productive working relationship. As Honorary Secretary, Fleure had begun the politically astute policy of securing distinguished public figures or persons eminent in another discipline to serve as officers or board members or trustees of the Geographical Association. Myres was active in all three capacities in the association, and could always be counted on to intervene on its behalf at a critical period.

In 1918, for example, the British Ministry of Education put forward a plan for the reorganization of secondary school studies which provided for three groups of advanced studies: (a) classics; (b) modern studies, including two foreign languages and British history and literature as major subjects; and (c) maths and science. Geography was treated not as one of the major subjects within the groups, but as an auxiliary to any of them. Myres and the Oxford historian Sir Charles Lucas persuaded the Board that geography might be approved as a so-called 'main subject' (that is, taught on its own) in any of the three groups.

A little later a committee on classics was formed to increase the representation of classics within this scheme, and invited the Geographical Association to send representatives to discuss geography and the classics. Again Myres was called upon, along with Fleure, to make the case for geography before a committee consisting of such luminaries as Alfred North Whitehead, Gilbert Murray and the biblical geographer George Adam Smith. In these and other ways Myres served to secure both an auxiliary and an independent position for geography in the British educational system, and for these and other services to geography he was awarded both the Geographical Association's Gold Medal, in 1942, and the Victoria Medal of the Royal Geographical Society in 1953.

In a session called 'The development of geography' at the annual meeting of the Geographical Association in 1943, H.J. Fleure began his remarks with these words: 'In reviewing geographical thought, let us not restrict ourselves to those who are called geographers.' As the case of John L. Myres, a classicist in geography, illustrates, skilled advocacy for geography, as well as its skilled professional practice, is not so restricted. Myres' example becomes relevant both as concerns the place of geography and as concerns the understanding of the geography of place, as a research concern and as an educational subject.

Bibliography and Sources

1. SELECTED BIBLIOGRAPHY OF WORKS ABOUT JOHN L. MYRES

Boardman, John, 'Sir John Linton Myres', *Dictionary of National Biography, 1951–1960*, London, Oxford University Press, 1971, 762–3.

Dunbabin, T.J., 'Sir John Myres', *Proceedings of the British Academy*, Vol. 41 (1955), 349–65.

Fleure, H.J., 'Sir John Linton Myres', *Geography*, Vol. 39 (1954), 128.

Gray, Dorothea, 'John Linton Myres', in Myres, *Homer and His Critics*, London, Routledge and Kegan Paul, 1958, 223–51.

K.M. [Kenneth Mason], 'John Linton Myres, 1869–1954', *Geogr. J.*, Vol. 120 (1954), 541–2.

Myres, J.N.L., *Commander J.L. Myres, R.N.V.R.: The Blackbeard of the Aegean*, London, Leopard's Head Press, 1980 (tenth J.L. Myres Memorial Lecture).

Myres manuscript material may be found in the archives of the institutions with which he was connected: Oxford University, the University of Liverpool, and the Royal Geographical Society. An unfinished autobiography remained with the family. Portraits may be found in New College and the School of Geography, Oxford.

2. SELECTED BIBLIOGRAPHY OF WORKS BY JOHN L. MYRES

1896   'An attempt to reconstruct the maps used by Herodotus', *Geogr.J.*, Vol. 8, 605–31.

1902   *A History of Rome*, London, Rivington.

1906   'The Alpine races in Europe', *Geogr. J.*, Vol. 28, 537–60.

1908   'Herodotus and anthropology', in R.R. Marett (ed.), *Anthropology and the Classics*, Oxford, Clarendon Press, 121–68.

1908   with C.M. Church, 'A contribution to the topography of north-western Greece', *Geogr. J.*, Vol. 32, 47–54.

1910   'The geographical study of Greek and Roman culture', *Scot. Geogr. Mag.*, Vol. 26, 113–30.

1910   *The Value of Ancient History*, Liverpool, Liverpool University Press.

1910   *Greek Lands and the Greek People*, Oxford, Clarendon Press (Inaugural Lecture).

1911   *The Dawn of History*, London, Williams and Norgate.

1911   'The geographical aspect of Greek colonization', *Proc. Class. Assoc.*, Vol. 8, 45–69.

1912   'Address to the research department, Royal Geographical Society', *Geogr. J.*, Vol. 40, 358–71.

1915–16   'The causes of rise and fall in the population of the ancient world', *Eugenics Review*, Vol. 7, 15–45.

1920   'The Dodecanese', *Geogr. J.*, Vol. 41, 329–47, 425–46.

1922   'The geographical distribution of the Greek city-states', *Geog. Teacher*, Vol. 11, 266–90.

1923   'Primitive man in geological times' and 'Neolithic and Bronze Age cultures', *Cambridge Ancient History*, Vol. I, 1–56, 57–111.

1924   'The Marmara region: a study in historical geography', *Scot. Geogr. Mag.*, Vol. 40, 129–50.

1925   'A geographical view of the historical method in ethnology', *Geog. Teacher*, Vol. 13, 9–29 (presidential address, Geographical Association).

1925–26   'Wayside geography', *Geog. Teacher*, Vol. 13, 285–95.

1926   'At the sign of the Mermaid: notes on a first cross between geology and history', *Geog. Teacher*, Vol. 13, 389–90.

1927   *The Political Ideas of the Greeks*, New York, Abingdon Press; London, Edward Arnold (Bennett Lectures, Wesleyan University).

1928   'Ancient geography in modern education', *Report of the British Association for the Advancement of Science* (Glasgow, 1928), 99–117 (presidential address, geography section, British Association).

1929   'The colonial expansion of Greece', *Cambridge Ancient History*, Vol. III, pp. 631–86.

1929   'Geography in relation to history and literature', *South African Geogr. J.*, Vol. 12, 23–5.

1930   *Who Were the Greeks?*, Berkeley, University of California Press (Sather Lectures).

1932   [Chair], 'What is historical geography?', *Geography*, Vol. 17, 39–45.

1934   'The ethnology and primitive culture of the Nearer East and the Mediterranean world' and 'The ethnology, habitat, linguistic and common culture of Indo-Europeans up to the time of the migrations', in G. Eyre (ed.), *European Civilization*, Oxford, Oxford University Press, Vol. I, 87–117, 184–244.

1936   'Region and race', *Geography*, Vol. 21, 18–27 (Herbertson Memorial Lecture).

1941   'The islands of the Aegean', *Geogr. J.*, Vol. 97, 137–57.

1941   'The position of Greece in the east Mediterranean', *Geography*, Vol. 31, 101–8.

1943   'The ancient shape of Attica', *Greece and Rome*, Vol. 12, 33–42.

1943   'The development of geography', *Geography*, Vol. 28, 71–5.

1944	*Mediterranean Culture*, Cambridge: Cambridge University Press (Frazer Lecture).
1949	'The geographical background of the Aegean civilization', *Archiv Orientalni*, Vol. 16, 196–204.
1953	*Geographical History in Greek Lands*, Oxford, Clarendon Press.
1953	*Herodotus: Father of History*, Oxford, Clarendon Press.
1958	*Homer and His Critics*, London, Routledge and Kegan Paul.

William A. Koelsch is Professor of History and Geography, Graduate School of Geography, Clark University, Worcester, MA 01610, USA.

Chronology

1869	Born on 3 July in Preston, Lancashire
1882–88	Student at Winchester School
1888–91	Undergraduate in New College, Oxford
1891	First visit to Greece
1892–94	Craven Fellowship; travel in Aegean
1892–95	Fellow of Magdalen College, Oxford
1894	First excavations on Cyprus
1895	Married Sophia Florence Ballance
1895–1907	Student in Christ Church, and University Lecturer in Classical Archaeology, Oxford
1901–03	Founding Editor, *Man* (also 1931–46)
1907–10	Gladstone Professor of Greek and Lecturer on Ancient Geography, University of Liverpool
1910–39	Wykeham Professor of Ancient History and Fellow of New College, Oxford
1911	Publication of *The Dawn of History*
1914	Sather Professor, University of California
1915–18	Wartime service (to lieutenant-commander, RNVR)
1919–32	General Secretary, British Association for the Advancement of Science
1920	D.Sc. (hon.), University of Wales
1923	Elected Fellow, British Academy
1925	President, Geographical Association
1927	Sather Professor, University of California
1928	President, Section E (Geography), British Association

1928–31	President, Royal Anthropological Institute
1929	D.Litt. (hon.), University of Witwatersrand
1930	Publication of *Who Were the Greeks?*
1933	D.Sc. (hon.), University of Manchester
1934–37	Chairman, British School at Athens
1935	Herbertson Memorial Lecturer
1935–38	President, Hellenic Society
1937	Ph.D. (hon.), University of Athens
1942	Gold Medal, Geographical Association
1943	Knighted by King George VI
1953	Victoria Medal, Royal Geographical Society; publication of *Geographical History in Greek Lands* and *Herodotus: Father of History*
1954	Died on 6 March in Oxford
1958	Publication of *Homer and His Critics*

Fridtjof Nansen

1861–1930

C.I. Jackson

Nansen's career had three distinct phases: scientist and explorer; Norwegian nationalist and diplomat; League of Nations humanitarian. It is in the first of these roles, especially the first crossing of Greenland and the Arctic Ocean drift of the *Fram*, that his contribution to geography is normally evaluated. From a human geography standpoint, however, what he achieved in the last dozen years of his life may be even more important. Philip Noel-Baker, who was well qualified to judge, declared that the true creators of the League of Nations were Woodrow Wilson, Lord Robert Cecil and Fridtjof Nansen. Nansen's humanitarian work for the League during the 1920s had lasting political and social consequences. Similarly, his role during the period in which Norway achieved full independence now appears significant in terms of the subsequent political geography of the northern countries. For the first three decades of the twentieth century, Nansen personified Norway in the eyes of the rest of the world; he therefore shaped both the world's perception of Norway, and Norway's own international role, in ways that are still evident.

Public renown and a lifetime of achievement concealed a troubled and brooding introspection. Infinitely careful in preparing for his pathbreaking explorations, and zealous in reporting on their scientific achievements, he remained a man of action, who began to plan the next venture as soon as the success of the present one had been assured. His marriage to Eva Sars was vitally important to him, yet they were separated by exploration or diplomacy for a substantial part of the eighteen years until her death in 1907. Brought up to have a strong sense of the need to maximize his talents through hard work, he became a difficult and demanding parent, as he himself recognized. The title of one biographical sketch of Nansen – 'Hamlet on Skis' (in Rasky's 1977 volume *The North Pole or Bust*) – is as inadequate as it is inelegant, yet the comparison to Hamlet was originally made by Nansen's daughter.

1. Education, Life and Work

Fridtjof Nansen's parents were Baldur Fridtjof Nansen (1817–75) and Adelaide Johanne Thekla Isidore Wedel Jarlsberg (1821–77), both of whom had children by previous marriages. Through his father, Nansen traced descent from the seventeenth-century Danish Arctic explorer and statesman Hans Nansen. His mother's family, the Wedel Jarlsbergs, was equally distinguished and it was at his mother's estate at Store Frøen, on the edge of Christiania (renamed Oslo in 1925), that Fridtjof Nansen was born on 10 October 1861.

Nansen's youth was characterized by a strong emphasis on outdoor activity, especially skiing. In 1880 he matriculated from the technical high school and in 1881 began studying zoology at the University of Christiania. It was as student and supernumerary zoologist on the sealer *Viking*, during the summer of 1882, that Nansen had his first experience of the east Greenland coast, and saw the massive volume of drift-ice and icebergs that sweeps southwards along the coast. This experience proved crucial in shaping his future career as explorer and scientist.

On his return from the *Viking* cruise, Nansen was appointed curator of the zoological collection in the Bergen Museum. There he remained for about seven years, contributing to the zoological literature, and refusing several employment offers from the United States. In 1885–6 he took a sabbatical in Italy, initially at Padua but mainly at the world's first biological station at Naples. This became the model for the station established at Drøbak in Christiania Fiord in 1894.

Nansen first envisaged crossing the central ice of Greenland in the autumn of 1883: 'It flashed on me like lightning: an expedition across Greenland from coast to coast on ski' (Sørensen, p. 44). It was clear to him that the crossing had to be from east to west, since the east coast was ice-bound and almost devoid of settlement. Going west 'one could always count on striking a settlement, and would consequently not need to provision for more than the anticipated duration of the crossing' (Sørensen, p. 45). A corollary of this logic, of course, was that there could be no turning back.

The Greenland expedition, four Norwegians and two Lapps, set out from Norway in early May 1888, immediately after Nansen had defended his doctoral thesis, 'The Structure and Combination of the Central Nervous System'. One of its members was Otto Sverdrup, nine years older than Nansen, who was to become captain of the *Fram*, and later an Arctic expedition leader himself. Sailing from Iceland in the sealer *Jason*, they were off Cape Dan (near the only significant east coast settlement at Ammassalik, 65° N) on 11 June, but the constant barrier of moving ice along the coast prevented them from reaching the shore. On 17 July they left the *Jason* in small boats, but were caught up in the ice stream and swept southwards. Not until 29 July, in latitude 61°40′, did they reach land; by a mixture of rowing and hauling the boats over the ice they were back in the area of Umivik Fiord (64° N) by 10 August.

The original plan had been to make a crossing from Cape Dan to Christianshåb on Disko Bay (68°50′ N). By 27 August Nansen realized that this was impracticable, because of the time that had been lost in reaching their starting point. The expedition headed instead for Godthåb (64° N). The highest point on the crossing was 2716 m (8911 ft). They reached ice-free land on the west coast on 24 September. The expedition had to remain in Godthåb throughout the winter, until the first ship reached the community in April 1889. They returned to Christiania on 30 May.

Later in 1889, Nansen married Eva Helene Sars (b. 1858), a romantic singer and the daughter of the zoologist Michael Sars. Their first child was born prematurely and did not survive. Their daughter Liv was born in early January 1893, several months before the beginning of the *Fram* expedition. Liv was subsequently one of her father's biographers (Høyer).

Nansen had conceived the idea for his next expedition as early as 1884. In June 1881, the De Long expedition's ship *Jeannette* had been crushed by pack ice north of the New Siberian Islands (longitude 155° E), after drifting northwestwards in the pack ice from Bering Strait for twenty-two months. No members of the expedition survived, but in 1884 items that could only have come from the *Jeannette* were found on the southwest coast of Greenland. In November 1884, the meteorologist Henrik Mohn speculated in a Norwegian newspaper that these relics must have drifted westwards across the polar basin until they entered the East Greenland Current and were swept around to the west coast. Nansen was not the only one to perceive that this drift could be utilized to explore the Arctic Ocean. In *Farthest North* he mentioned a similar suggestion from a Dane living at Julianhåb in Greenland. Only Nansen, however, was prepared to put the idea to the test.

Crucial to the success of the venture was the design of the ship. As Nansen wrote in 1891:

> I propose to have a ship built, as small and as strong as possible; just big enough to contain supplies of coals and provisions for twelve men for five years . . .
>
> The main point in this vessel is that it be built on such principles as to enable it to withstand the pressure of the ice. The sides must slope sufficiently to prevent the ice, when it presses together, from getting firm hold of the hull, as was the case with the *Jeannette* and other vessels. Instead of nipping the ship, the ice must raise it up out of the water. (Quoted in *Farthest North*, Vol. 1, p. 30).

Designed by Colin Archer, the *Fram* was launched on 26 October 1892. The expedition sailed from Christiania Fiord on midsummer day 1893, and left Vardø in northern Norway on 21 July. On 22 September, in the Laptev Sea at 77°43′ N, 134° E, the *Fram* ceased independent movement and began its drift in the pack ice.

The ship became a floating scientific station, on which the expedition members contrasted their health and comfort with the sufferings of earlier expeditions:

> I am almost ashamed of the life we lead, with none of those darkly painted sufferings of the long winter night which are indispensable to a properly exciting Arctic expedition. We shall have nothing to write about when we get home. (28 December 1893, *Farthest North*, Vol. 1, p. 313).

It was, however, precisely this tranquil existence, in which the *Fram's* journey was determined entirely by natural forces, that made Nansen increasingly introspective and impatient. The experience that he had envisaged and planned for so meticulously was one that he found difficult to endure. In his diaries, and especially the lengthy entry for 19 June 1894 (quoted in Sørensen, pp. 138–46), there are soliloquies that are reminiscent of Hamlet. But there is also a vital contrast. Hamlet berates himself for his inability to act, whereas Nansen broods because his natural inclination to action has been constrained. When Hamlet finally acts decisively, killing Polonius and setting in motion the ultimate tragedy, it is a spur-of-the-moment reaction, essentially out of character. Nansen, however, usually based his action on careful planning that recognized the dangers involved. As he wrote in March 1894:

> I know this is all a morbid mood; but still this inactive, lifeless monotony, without any change, wrings one's very soul. No struggle, no possibility of struggle! All is so still and dead, so stiff and shrunken under the mantle of ice. Ah! . . . the very soul freezes. What would I not give for a single day of struggle – for even a moment of danger. (*Farthest North*, Vol. 1, pp. 379–80; a slightly different version appears in Sørensen, p. 134)

Though the circumstances of the *Fram*'s drift gave full scope for Nansen's gloomy introspection, it was a characteristic that appeared many times in the remainder of his life. It was severe at this time because, although Nansen had anticipated the drift might take three years, and had provisioned for five, the movement of the ice suggested to him that the *Fram* might drift for seven years before being released. By the spring of 1894, therefore, less than a year into the drift, Nansen had already begun to plan his escape back to action. It was becoming evident that the *Fram*'s drift would not pass close to the North Pole (it reached 85°55′30″ N in November 1895); why not then use the *Fram* as a launch pad for an assault on the Pole by men on skis?

The principal objection, of course, was that it would not be possible to return to the *Fram*, since its location would be unknown to the ski party. This, however, was not an objection that weighed hard with someone who had crossed Greenland without the possibility of return.

By late in 1894 Nansen was determined to make the attempt; he and Hjalmar Johansen left the *Fram* on 14 March 1895, in latitude 84° N, longitude 102° E. What has become known as 'the fifteen months' sledge expedition' (from the subtitle of *Farthest North*) was in reality something rather less. Their journey did last fifteen months, and sledges (used normally and also converted into makeshift kayaks) provided the means of travel. But the attempt to reach the North Pole was abandoned because of ice conditions on 8 April, less than a month after leaving the *Fram*. For most of the fifteen months, from 26 August 1895 until 19 May 1896, Nansen and Johansen wintered in a makeshift hut on the coast of Franz Josef Land, with nothing to read but navigation tables and with no accurate knowledge of their position, since they had allowed their watches to run down.

Had they achieved the Pole, their hardships and heroism would of course appear in a different light. Nansen and Johansen had achieved the 'farthest north' at 86°14′ N, but it is difficult to avoid the conclusion that, for once, Nansen's unique blend of careful planning and courageous action had been misplaced.

Moving southwestwards along the shores of Franz Josef Land during the early summer of 1895, Nansen and Johansen encountered another Arctic expedition, led by Frederick Jackson, on 17 June. They remained at Jackson's base for a month, alternately rejoicing in the relative comfort and opportunity for scientific discussion ('I felt like a piece of parched soil drinking in rain after a drought of a whole year': *Farthest North*, Vol. 2, p. 488) and equally characteristically becoming impatient for the arrival of Jackson's relief ship ('We regretted that we had not at once pushed on for Spitzbergen', *Farthest North*, Vol. 2, p. 490). The *Windward* did arrive on 26 July and sailed on 7 August. Nansen and Johansen reached Vardø on 13 August 1896. The *Fram*, released from the ice in the three years that Nansen had originally envisaged, arrived at Skjervøy only one week later. Reunited with his wife in Hammerfest, and with the rest of the expedition in Tromsø, Nansen returned to Christiania Fiord in the *Fram* on 9 September.

Characteristically, Nansen came back with plans for a new expedition, using the *Fram* as a basis for Antarctic research. As with the Greenland and Arctic Ocean expeditions, these ideas had to be deferred for several years. Nansen was now appointed a research professor at the University of Christiania. *Farthest North*, the two-volume popular account of the expedition, was published in 1897; it was followed between 1900 and 1905 by six volumes of scientific results, with Nansen as general editor and author of several parts. At the same time, Nansen and Norway were strengthening their commitment to oceanographic research. Nansen joined the first cruise of the research vessel *Michael Sars*, named for his father-in-law, in the Norwegian Sea in 1900, a cruise that generated further scientific reports.

Nansen wrote in November 1902 that:

> I am always travelling towards something vague which I never reach. What I am engaged on, whether it be an expedition across the Arctic or work on a scientific problem, is and will always be something temporary, something that must be cleared out of the way first in order to get to the reality which is life. And so it will go on until one day I drop . . . The reality has never been attained. (Quoted in Sørensen, p. 215)

What got in the way in 1905, and changed decisively the course of Nansen's life, was the crisis with Sweden that resulted in full Norwegian independence. This independence had nominally existed since 1814, though the two nations had been linked together in the person of a single monarch. Norway, however, had always been the junior partner. From 1885 onwards, divergent views on the nature of the constitutional relationship, especially in regard to Norway's external relations, led to a situation in which war between the two countries became a real possibility.

Nansen played a crucial role during the crisis in several ways. His international reputation made him a powerful and credible voice, heard by both governments and their publics, arguing either for greater Norwegian freedom of action within the Union, or for independence if this was not possible. His was also a voice of reason, making it clear that, once the constitutional problems were resolved, Norway had no quarrel with Sweden. Thirdly, he helped to ensure that Norway would become a constitutional monarchy rather than a republic, a matter of some importance to other European powers. He was Norway's chosen emissary to Prince Carl of Denmark, who became King Haakon VII following decisive plebiscites in favour of both independence and a monarchical system.

Immediately after independence was achieved, Nansen became Norway's ambassador to Britain, with the task of negotiating the Integrity Treaty, signed in 1907. The diplomatic life in Edwardian London was not easily reconciled with either Nansen's family life in Norway (where Eva and the children remained during most of this period) or his continuing desire to use the *Fram* in an effort to reach the South Pole. In September 1907, Nansen relinquished his Antarctic ambitions by promising Amundsen the use of the *Fram* for what was originally intended as a new drifting expedition in the Arctic Ocean via Bering Strait, and subsequently became Amundsen's own successful South Pole expedition.

In December 1907, Eva Nansen died unexpectedly, and Nansen's family misfortunes increased in 1913, with the death of his young son Åsmund of cerebral meningitis. There remained, however, the scientific work. Nansen's position at the University of Christiania had been formally changed to a chair in oceanography, and the pattern of cruises and subsequent research papers might have continued indefinitely but for the outbreak of war in Europe in 1914.

In neutral Norway, Nansen was well able to appreciate the implications of the First World War. After the USA joined the war in 1917, Nansen's prestige and diplomatic skills were needed in Washington, to ensure continuation of Norwegian food imports. But Nansen's preoccupation in his diaries was with wider issues, and this time his soliloquies seem unarguable:

> What a nightmare of insanity; and no one can stop it – no one. The people of Europe, 'the torch-bearers of civilisation.' are devouring one another, trampling civilisation under foot, laying Europe in ruins; and who will be the gainer? . . . A civilization that sets up power as its aim and ideal cannot possibly achieve progress for humanity. It must inevitably lead hither – towards destruction . . .
>
> There must be a regeneration – a new era with new ideals – when spiritual values will again be the end and material values only a means, when the world will no longer be ruled by mediocrity and the mob. (Quoted in Sørensen, pp. 271, 273)

The League of Nations seemed the best hope for such regeneration. Nansen was influential in Norway's accession to the League, and he was the chief Norwegian delegate to the League's sessions during the 1920s. In a broader context, he worked to ensure that the League included smaller and neutral countries, instead of being only a meeting place of the Great Powers, and he argued strongly and ultimately successfully for the admission of Germany to the League.

It was, however, as the League's Commissioner for Refugees that Nansen's name became known to hundreds of thousands of people who knew little or nothing of his Arctic achievements. As was to happen again in 1945, the end of the First World War left vast numbers of prisoners and refugees needing repatriation. The situation was further complicated by the Russian Revolution and the continuing hostilities and distrust between the Soviet government and other countries. With the International Red Cross as Nansen's main executive arm, 'during 1920 and 1921, about 437,000 prisoners were liberated . . . and brought home to their own countries' (Sørensen, p. 281).

No sooner had the prisoners of war been repatriated than Nansen had to turn to the more intractable problem of a million and a half Russian refugees. Only six thousand returned to Russia. Though many were resettled during the 1920s, at the end of the decade there were still 'over a million refugees in Europe for whom no definite arrangement had been made' (Sørensen, p. 287). The assignment that Nansen accepted in 1920, nominally only for two months, has evolved into the never-ending task of the United Nations High Commissioner for Refugees.

A third strand in Nansen's humanitarian work during the last decade of his life was his attempt to relieve famine in Russia, and the associated disease and death involving millions of people. He was supported strongly by Herbert Hoover in the USA and by private funds from many countries; more ironically, Vidkun Quisling was his principal colleague and administrator. But Nansen saw, in the unwillingness of the Great Powers to distinguish humanitarian need from their fear and detestation of the Soviet Union, that the League would not fulfil the ideals on which it was founded.

In the midst of his humanitarian burdens, Nansen still found some time for science. A series of papers throughout the 1920s indicates his continuing interest in the earth sciences and particularly Arctic research and exploration.

Fridtjof Nansen died at his home near Oslo on 13 May 1930 at the age of 69. He is commemorated in place names in Canada, Greenland and Antarctica; however, the two principal memorials to his life and work are the *Fram*, on permanent exhibit in Oslo, and the Office of the United Nations High Commissioner for Refugees.

2. Scientific Ideas and Geographical Thought

In Fridtjof Nansen, adventurous heart and logical and scientific head were normally reconciled and mutually supportive. Although the Greenland crossing was more an adventure than a scientific enterprise, discovering what lay behind the coastal glaciers and ice sheet of west and east Greenland was important for the development of both earth and atmospheric sciences in the late nineteenth century. Nordenskiöld, who had failed in his own attempt to reach the interior in 1870, supported Nansen with the statement that 'the investigation of the real nature of Greenland is of such great and fundamental significance for science that it is scarcely possible at the present time to set a more important goal for a polar expedition'. Although the expedition was a fast-moving one, which had little time to take more than simple measurements of location, altitude and weather, these were sufficient to confirm that Greenland consisted of a massive ice sheet, rising to thousands of feet above sea level. There were no ice-free, potentially inhabitable, valleys beyond the coastal nunataks, as some had imagined.

Nansen took several years to plan his major expeditions, across Greenland and in the *Fram*. He was ready to change these plans at short notice when necessary, as exemplified in the *en route* decision to aim at Godthåb rather than Christianshåb during the Greenland crossing. Nansen was also capable of making use of unexpected opportunities; his book *Eskimo Life* was a major contribution to anthropology and human geography, for which the need to winter at Godthåb provided the opportunity, and made possible by Nansen's ability to learn the language and to gain the confidence of his Eskimo hosts.

It was, however, the *Fram* expedition that represented an opportunity that few others would have recognized, and fewer still would have been prepared to undertake. Virtually no one else would have had the capacity to bring the scientific, shipbuilding and other skills together so triumphantly. Though its principal origins are the *Jeannette* relics from 1881, found three years later in southwest Greenland, it may be that Nansen's zoological training had already made him receptive to such a discovery. From his earliest (1882) voyage to the east Greenland coast in the *Viking*, he had brought back from the ice floes samples of organic slime ('exuviae of microscopic silica algae'). Of the sixteen species identified, twelve were known only from Siberian seas. 'Nansen drew the conclusion that there was some connection between the sea near the Bering Strait and that off the coast of Greenland' (Høyer, p. 17; see also *Farthest North*, Vol. 1, p. 24).

After his Greenland crossing, Nansen had become an international figure. Nevertheless, his plans to build a ship that could survive the Arctic pack ice and follow the route presumably travelled by the *Jeannette* relics invited criticism that was similar to that which greeted his Greenland proposal. The voyage was an affront to conventional thinking, and most of the Arctic establishment in England was convinced that it was impracticable. From the United States, A.W. Greely concluded a devastating critique as follows: 'Arctic exploration is sufficiently credited with rashness and danger in its legitimate and sanctioned methods, without bearing the burden of Dr. Nansen's illogical scheme of self-destruction' (quoted in *Farthest North*, Vol. 1, p. 52).

In the opening chapter of *Farthest North*, Nansen reprinted, presumably with well-concealed amusement, many of the criticisms 'proving' why the expedition could not possibly succeed. Some argued that the identification of the *Jeannette* relics was erroneous, others that there was no ship design capable of withstanding the polar pack, and yet others that there was a strong probability of land in high latitudes: a few small relics might pass through the intervening channels, but a ship would probably be wrecked. Of all Nansen's critics, George Nares was among the best qualified to judge, having taken HMS *Alert* to the north coast of Ellesmere Island (82°30′ N) and brought her back in 1875–76. According to Nares,

> Supposing the sea currents are as stated, the time calculated as necessary to drift with the pack across the polar pack is several years, during which time, unless new lands are met with, the ice near the vessel will certainly never be quiet, and the ship herself never free from the danger of being crushed by ice presses. To guard against this the vessel is said to be unusually strong, and of a special form to enable her to rise when the ice presses against her sides. This idea is no novelty whatever; but when once frozen into the polar pack the form of the vessel goes for nothing. She is hermetically sealed to and forms a part of the ice block surrounding her. The form of the ship is for all practical purposes the form of the block of ice in which she is frozen. This is a matter of the first importance, for there is no record of a vessel frozen into the polar pack having been disconnected from the ice, and so rendered capable of rising under pressure as a separate body detached from the ice block, even in the height of summer. (Quoted in *Farthest North*, Vol. 1, pp. 42–3)

Nares also doubted that the drift would carry the *Fram* into high latitudes.

Nansen was right and Nares was wrong, and Nares was subsequently gracious in acknowledging this. But, on at least one occasion during the drift, it was a close thing. Early in January 1895, it seemed as though the *Fram* was doomed. The expedition was ready to evacuate her ('We are now living in marching order on an empty ship': *Farthest North*, Vol. 2, p. 52) and, when the danger was passed, Nansen admitted that

> Had this attack on the *Fram* been planned by the aid of all the wickedness in the world, it could not have been a worse one. The floe, 7 feet thick,

has borne down upon us on the port side, forcing itself up on the ice in which we are lying, and crushing it down. Thus, the *Fram* was forced down with the ice, while the other floe, packed up on the ice beneath, bore down on her, and took her amidship while she was frozen fast . . . Had the *Fram* not been designed as she was, we should not have been sitting here now. Not a drop of water is to be found in her anywhere. (*Farthest North*, Vol. 2, pp. 55–6)

This was a rare event: for most of the three years, the ship was, as Nansen had planned, a tranquil oceanographic research station, exploring an unknown ocean at a time when the scientific study of the oceans, as well as of sea ice and the polar atmosphere, was still in its infancy. This being so, it seems surprising that the expedition members contained no recognised scientist other than Nansen himself. In *Farthest North*, Nansen gave little explanation of the reasons why most of his twelve colleagues were selected, apart from their volunteering and being in good health. According to the Curator of Manuscripts at the University of Oslo,

Nansen had hoped to have a crew consisting of men with higher education who could combine scientific work with the work of ordinary crew members . . . The plan was not feasible. No one could know how many years the drift would last and therefore Nansen was not in a position to offer fixed wages for the duration. He could not get trained scientists to sign up on those conditions and therefore had to do without them. (Vasstveit, personal communication, 1994)

Whether viewed in heroic or scientific terms, the expedition was, however, immensely successful. A.W. Greely was almost alone in criticising Nansen for embarking on the sledge expedition, and Greely's objections (published only a few days after the expedition's return) were clearly misconceived. He accused Nansen of 'deliberately quit[ting] his comrades on the ice-beset ship hundreds of miles from any known land' and thereby deviating 'from the most sacred duty devolving on the commander of a naval expedition' (quoted in *Farthest North*, Vol. 1, pp. 52–3). As indicated earlier, the sledge journey can be criticised, but scarcely for that reason.

3. Influence and Spread of Ideas

A. NANSEN AS EXPLORER-SCIENTIST

The first crossing of Greenland in 1888 was significant in three distinct respects. First, and most directly, it confirmed the general model of the interior of Greenland held by most scientists: a vast and continuous mass of ice. Other crossings were made in the period up to 1914, but detailed glaciological investigations of the inland ice had to wait until after the Second World War. Nevertheless, confirmation of the general nature of the Greenland ice was a powerful stimulus to understanding northern hemisphere meteorology, and ensured that Greenland would be a key element in theories of climate and climate change (such as the 'glacial anticyclone' theory of W.H. Hobbs).

Secondly, the circumstances of the Greenland crossing produced an unexpected but major contribution to anthropology and human geography. The crossing took place much later in the year than had been anticipated, requiring the expedition to spend a long winter at Godthåb awaiting the 1889 navigation season. Nansen seized this opportunity to learn the Eskimo language and to establish a rapport with the local community that enabled him to produce his volume *Eskimo Life*.

Lastly, the Greenland expedition established Nansen's credentials as explorer and scientist. Greely grumbled, prior to the *Fram* expedition, that 'Dr. Nansen, so far as I know, has had no Arctic service; his crossing of Greenland, however difficult, is no more polar work than the scaling of Mount St. Elias' (quoted in *Farthest North*, Vol. 1, p. 49). This, however, was a minority view. Most Arctic scientists were sceptical about the *Fram* expedition proposals, and some found them alarming, but his new proposals were taken very seriously.

The three-year drift of the *Fram* was as successful in scientific terms as it was in vindicating Nansen's belief that a specially designed ship could safely retrace the journey taken by the *Jeannette* relics. The *Fram* became a floating scientific station that, from an oceanographic standpoint, combined the virtues of a fixed site (for such tasks as depth measurements and deep-water sampling) with the slow drift that was vital to understanding the relationship between wind, ice and surface currents. In shaping the emerging science of oceanography, the drift of the *Fram* was of major importance.

It was the relationship of wind stress to surface current movement that was the principal result of the *Fram* expedition, because of its general significance for the world oceans. What is now known as the 'Ekman spiral' – the changes of current direction and speed at increasing depth caused by a combination of surface wind and Coriolis force – was in fact suggested by Nansen as a result of his analysis of the *Fram* drift, though it was Ekman who gave this suggestion its precise mathematical formulation in 1905.

The second major finding of the expedition was the unexpected depth of the Arctic Ocean. As Nansen wrote at the conclusion of the expedition,

> [W]e have demonstrated that the sea in the immediate neighbourhood of the Pole, and in which, in my opinion, the Pole itself in all probability lies, is a deep basin, not a shallow one containing many expanses of land and islands, as people [Nansen included] were formerly inclined to assume. (*Farthest North*, Vol. 2, p. 631)

The most direct evidence for the existence of the deep Angara Basin was provided by the soundings that the *Fram* expedition was able to make, but Nansen realized that the unimpeded movement of the pack ice in a northward direction, and the immense volume of drift ice in the Greenland Current, were indications that the area of the polar basin was liable to be substantial.

Another unexpected finding was the thermohaline profile:

> [I]t was customary to look upon the polar basin as being filled with cold water, the temperature of which stood somewhere about $-1.5°C$. Consequently our observations showing that under the cold surface there was warmer water, sometimes at a temperature as high as $+1°C$, were surprising. Again, this water was more briny than the water of the polar basin has been assumed to be. (*Farthest North*, Vol. 2, p. 634)

The *Fram* expedition turned Nansen's scientific interests decisively towards physical oceanography, and Norway became a leader in the field. The first decade of the twentieth century has been described as the 'golden age' of oceanography. The International Council for Ocean Research was created in 1902, and the International Oceanographic Laboratory was established at Christiania. From his experience in the *Fram* drift, Nansen devised, among other scientific devices, the Nansen bottle, for many years used to obtain water samples, and associated temperatures, at depth.

B. NANSEN AS NORWEGIAN NATIONALIST

Nansen played as great a role as anyone in shaping the political geography of Norway during the twentieth century. This concern long predated his specific and crucial role in the events of 1905: as he recounted in *Farthest North*, when he encountered Frederick Jackson in Franz Josef Land in June 1895, his first questions were about his wife and daughter; 'Then came Norway's turn, and Norwegian politics.'

Significantly, Jackson was able to provide a response to the family questions, but knew nothing of Norwegian affairs. Nansen's first role during the constitutional crisis a decade later was to use his international fame as a means of making Norway's predicament known and understood outside Scandinavia, and especially by the public and governments of Britain and Germany. Initially these Great Powers favoured the status quo, which meant support of the Swedish position; their acceptance of full Norwegian independence was due in no small part to Nansen's ability to persuade them that this was a solution of a problem, not the creation of new problems. Nansen's own personality was a vital element, enabling him to establish a personal relationship with Edward VII and other national leaders, and appearing as Arctic hero to a wider public.

These personal characteristics were required again a decade later, when the entry of the United States into the First World War put Norwegian imports of wheat and other vital foodstuffs in peril. As seen from Washington, Norway was a neutral nation with close economic and other ties to Germany; it required several months of patient diplomacy in Washington to obtain American recognition of Norway's need for exemption from wartime US export controls.

C. NANSEN AS INTERNATIONALIST

There is no real contradiction between Nansen's long and intense support for Norwegian independence, and his equally strong conviction that the First World War demonstrated that a new era of international co-operation was necessary. Nansen's internationalism did not minimize the nation state, or anticipate its decline; the organisation he helped to create was indeed a League of Nations. As he had anticipated, the achievement of Norwegian independence resolved the longstanding dispute with Sweden, and the subsequent history of the northern countries exemplifies Nansen's brand of internationalism: nation states should be the means of achieving international co-operation, not barriers to it.

Though the League proved to be inadequate, this was not obvious at its inception. It was Nansen, more than any other individual, who showed that under the League's auspices great things could be accomplished.

Nansen's internationalism had already been recognized by his appointment, at the end of the war, as the neutral chair of an arbitration commission to settle disputes between Britain and the United States (Sørensen, p. 275). And it was as Nansen himself, as much as the League's Commissioner, that he succeeded in the task of repatriation. The Soviet government would deal with him personally, but not as the League's representative. He made possible, in the name of the League, 'an international loan for countries unable to secure credit for themselves' (Sørensen, p. 279), but he also raised much of the money needed for repatriation through an international non-governmental organization, Nansen Relief. Out of this period came the 'Nansen passport':

> One problem was that the refugees had neither passports nor papers which other states would recognize. Nansen called together representatives from the various governments to a meeting in Geneva in July 1922. Thirty-one were represented, and they accepted Nansen's proposal for an identification certificate for each refugee which could be used as a passport. Fifty-two governments have recognized these certificates, which are stamped with Nansen's picture and known as 'Nansen Passports.' . . . Up to 1930 there have been several hundred thousand of these issued. (Sørensen, p. 285)

Within a decade of its foundation, the inadequacy of the League became evident. Nevertheless, the name 'Nansen' could open doors in the Kremlin and cheque-books in many countries, and the challenges he faced during the 1920s made full use, perhaps for the first time, of all the abilities that he had exhibited earlier in his career. He could define needs quickly and clearly; his ingenuity, allied to his planning and organizational skill, turned needs into workable solutions; and his ability to communicate and persuade broke diplomatic barriers and generated sources of funding. The problems of humanity during the 1920s left little time for brooding introspection; impatience became a virtue instead of a burden.

Nansen was awarded the Nobel Peace Prize in 1922, but it was not until a year after his death that his humanitarian achievements could be summarized by the League of Nations:

> Dr Fridtjof Nansen acted for ten years as High Commissioner of the League of Nations, first for the repatriation of prisoners of war, and later for the protection and assistance of refugees . . . By his unwearied labor, and thanks to his courage, his perseverance, and his organizing power, he repatriated nearly half a million prisoners of war belonging to more than thirty different nations, and helped to make tolerable the lot of over 1,250,000 Greek, 1,000,000 Russian, 300,000 Armenian and some tens of thousands of Assyrian, Assyro-Chaldean, Bulgarian and Turkish refugees.
>
> In bringing about these results, Dr Nansen . . . travelled to almost every country in Europe, several times to Russia and to America: he made prolonged expeditions to Constantinople and to Greece after the events of 1922, and to the Armenian Republic of Erivan in 1925. He was thus compelled for long periods virtually to abandon other work, while his unceasing efforts did much to undermine his health and strength. (Quoted in Sørensen, p. 335)

Bibliography and Sources

1. SELECTED BIOGRAPHICAL AND AUTOBIOGRAPHICAL SOURCES

1896   Brögger, W.C. and Nordahl Rolfsen, *Fridtjof Nansen*, trans. William Archer, London, Longmans, Green and Co., 402 pp.

1899   Johansen, Hjalmar, *With Nansen in the North*, London, Longmans, Green and Co., 351 pp.

1932   Sørensen, Jon, *The Saga of Fridtjof Nansen*, trans. J.B.C. Watkins, New York, W.W. Norton; London, George Allen and Unwin, 372 pp.

1943   Sonntag, Wolfgang, *Held des Friedens, Fridtjof Nansens Leben*, Zurich, Buchergilde Gutenberg, 424 pp. A Norwegian translation by Nansen's daughter Liv Nansen Høyer, with additional chapters by the translator, was published in Oslo by Gylendal in 1947 (468 pp.).

1957   Høyer, Liv Nansen, *Nansen – A Family Portrait*, trans. Maurice Michael, London, Longmans, Green and Co., 269 pp.

1959   Shackleton, Edward A.A., *Nansen, the Explorer*, London, H.F. and G. Witherby, 209 pp.

1962   Noel-Baker, Philip, *Nansen's Place in History*, Oslo, Universitetsforlaget, 26 pp. Concerned only with Nansen's international role from 1920 onwards.

1973   Greve, Tim, *Fridtjof Nansen 1861–1904*, Oslo, Gyldendal, 261 pp. In Norwegian.

1981   Bauer, Walter, *Fridtjof Nansen. Humanität als Abenteuer*, Frankfurt, Fischer, 263 pp.

1991   Brux, Karl H., *Eva og Fridtjof Nansen, Et Samliv*, Oslo, Gyldendal, 366 pp. In Norwegian.

2. WORKS AUTHORED OR SUBSTANTIVELY EDITED BY FRIDTJOF NANSEN

Because many of Nansen's works appeared originally in Norwegian (less frequently in German or other languages), this listing cites the original publication and, where relevant, the appearance of an English-language version.

1884   'Langs Grønlands østkyst' ['Along the east coast of Greenland'], *Geografisk tidskrift*, Vol. 7, 76–9.

1885   'Bidrag til myzoostomernes anatomi og histology' ['On the anatomy and histology of the Myzostomidae'], Bergen Museum, 86 pp.

1886   *The Structure and Combination of the Histological Elements of the Central Nervous System*, Bergen Museum.

1887   Several zoological papers, in Norwegian, in *Naturen*, and also a study of the hagfish: 'A protandric hermaphrodite (*Myxine glutinosa*) amongst the vertebrates', Bergen Museum. Also Nansen's doctoral thesis: 'Nerveelementer, deres struktur og sammenheng i central-nervesystemet'.

1889   'Journey across the inland ice of Greenland from east to west', *Proceedings of the Royal Geographical Society*, Vol. 11, 469–87.

1890   *Paa ski over Grønland* . . ., Christiania, Aschehoug, 704 pp. English translation by H.M. Gepp, with additional material, published as *The First Crossing of Greenland*, London, Longmans, Green and Co., 1890, 2 vols, 510 pp and 509 pp.

1891   *Eskimoliv*, Christiania, Aschehoug, 263 pp. English translation by William Archer published as *Eskimo Life*, London, Longmans, Green and Co., 1893, 350pp.

1892   ed., with Henrik Mohn, *Wissenschaftliche Ergebnisse von dr. F. Nansens Durchquerung von Grönland 1888* [*Scientific Reports of Dr F. Nansen's Crossing of Greenland, 1888*], Gotha, Perthes, 111 pp. Nansen himself wrote the section on geology and hydrography (pp. 53–103).

1893   'How can the north polar regions be crossed?' *Geographical Journal*, Vol. 1, 1–32.

1894   with Gustav Guldberg, *On the Development and Structure of the Whale. Part I.* Bergen, John Grieg.

1897   *Fram over polhavet, den norske polarfaerd 1893–96*, Christiania, Aschehoug, 2 vols. English translation published as *Farthest North*, London, Constable, 1897, 2 vols, 510 pp. and 671 pp.

1898   'Future north polar exploration', *McClure's Magazine*, Vol. 10, No. 4, 293–305.

1900–06   ed., *The Norwegian North Polar Expedition 1893–1896, Scientific Results*, London and New York, Longmans; Christiania, Dybwad, 6 vols. Includes Nansen's own contributions: (a) 'The oceanography of the north polar basin', Vol. 3, No. 9, 427 pp.; (b) 'Protozoa on the ice-floes of the north polar sea', Vol. 5, No. 16, 22 pp.; (c) 'The bathymetrical features of the north polar seas with a discussion of the continental shelves and previous oscillations of the shore-line', Vol. 4, No. 13, 231 pp.; (d) 'On hydrometers and the surface tension of liquids', Vol. 3, No. 10; (e) with Robert Collett, 'An account of the birds', Vol. 1, No. 4, 53 pp.

1901   'Some oceanographical results of the expedition with the *Michael Sars* in the summer of 1900. Preliminary report', *Nytt magasin for naturvidenskapene*, Vol. 39, No. 2, 129–61.

1904   'Was ist Totwasser?' ['What is dead water?'], *Annalen der Hydrographie und maritimen Meteorologie*, Vol. 32, 309–10.

1905   'Die Ursachen der Meeresströmungen' ['The causes of ocean currents'], *Petermanns geographische Mitteilungen*, Vol. 51, No. 1, 1–4; No. 2, 25–31.

1905   *Norway and the Union with Sweden*, London, Macmillan, 96 pp.

1906   *Northern Waters: Captain Roald Amundsen's Oceanographic Observations in the Arctic Seas in 1901*, Christiania, Dybwad, 145 pp.

1907   'On north polar problems', *Geographical Journal*, Vol. 30, 469–87, 585–601.

1909   with Bjørn Helland-Hansen, *The Norwegian Sea, Its Physical Ocean-ography, based upon the Norwegian Researches 1900–1904*, Christiania, Mallingske Bogtrykkeri, 390 pp.

1911   *Nord i taakeheimen: utforskningen av jordens nordlige strøk i tidlige tider*, Kristiania, Dybwad, 603 pp. English translation by Arthur G. Chater published as *In Northern Mists: Arctic Exploration in Early Times*, London, Heinemann, 1911, 2 vols, 383 pp. and 415 pp.

1911   'The Norsemen in America', *Geographical Journal*, Vol. 38, 557–80.

1912   'Das Bodenwasser und die Abkühlung des Meeres' ['Ocean-bottom water and cooling of the sea'], *Internationale Revue der gesammten Hydrobiologie und Hydrographie*, Vol. 5, No. 1, 1–42.

1912   with Bjørn Helland-Hansen, *The Sea West of Spitsbergen: The Oceanographic Observations of the Isachsen Spitsbergen Expedition in 1910*, Christiania, Jacob Dybwads, 89 pp.

1912   with Bjørn Helland-Hansen, *The Oceanographic Investigations with the Fram in the North Atlantic 1910 and the South Atlantic 1911*, Christiania, Jacob Dybwads.

1914   with Jonas Lied, 'The sea-route to Siberia', *Geographical Journal*, 481–500. Nansen's section is entitled 'On the ice conditions and the possibility of a regular navigation'.

1914   *Gjennen Sibirien*, Christiania, Jacob Dybwads, 386 pp. English translation by Arthur G. Chater published as *Through Siberia, the Land of the Future*, London, Heinemann, 1914, 478 pp.

1915   *Spitsbergen Waters: Oceanographic Observations during the Cruise of the* Veslemöy *to Spitsbergen in 1912*, Christiania, Dybwad, 132 pp.

1917   with Bjørn Helland-Hansen, *Temperatur-Schwankungen des Nord-atlantischen Ozeans und in der Atmosphäre*, Christiania, Dybwad, 341 pp. English translation, with additions, published as *Temperature Variations in the North Atlantic Ocean and in the Atmosphere: Introductory Studies on the Cause of Climatological Variations*, Washington, Smithsonian Institution, 1920, 408 pp.

1920   'Spitsbergens opdagelse' ['The discovery of Spitsbergen'], *Naturen*, Vol. 44, Nos 1–2, 1–12.

1922   *The Strandflat and Isostasy*, Christiania, Dybwad, 313 pp.

1923   *Russland of Freden*, Christiania, Dybwad. English translation published as *Russia and Peace*, London, Allen and Unwin, 1923.

1925   *Hunting and Adventure in the Arctic*, New York, Duffield, 462 pp. This translation of *Blandt sel of bjørn* (Christiania, 1924) is an account of Nansen's first trip to the Arctic in the sealer *Viking* in 1882.

1925   *Klimat-vekslinger i Nordens historie* [*Climatic Changes in the History of the North*], Oslo, Dybwad, 63 pp.

1926   *Klima vekslinger i historisk og postglacial tid* [*Climatic Changes during Historical and Post-glacial Times*], Oslo, Dybwad, 26 pp. This and the preceding item form part of a controversy with others over evidence for major climate changes.

1927   *Adventure, and Other Papers [by] Fridtjof Nansen*, London, L. and Virginia Woolf, 81 pp.

1927   'Svalbard', *Norsk geografisk tidsskrift*, Vol. 1, 10–30.

1927   'The scientific importance of Arctic research', published in the *Verhandlungen* [Proceedings] of the first regular meeting (Berlin 1926) of the International Society for the Exploration of the Arctic Regions by Means of Aircraft, Gotha, Perthes, 115 pp. The proceedings, including Nansen's paper, were published in German. Nansen subsequently became the founding editor (1928–29) of the Society's journal *Arktis*.

1928   *Gjennem Armenia*, Oslo, Dybwad. English translation published as *Armenia and the Near East*, London, Allen & Unwin, 1928.

1928   'The oceanographic problems of the still unknown Arctic regions', in American Geographical Society, *Problems of Polar Research*, New York, American Geographical Society, 2–4.

1928   *Die Gleichgewichtbewegungen der Erdkruste und die Oberflächen der Kontinente* [*The Earth's Crust, Its Surface Forms, and Isostatic Adjustment*], Oslo, Jacob Dybwads.

1929   *Gjennem Kaukasus til Volga*, Oslo, Dybwad. English translation published as *Through the Caucasus to the Volga*, New York, Norton.

3. PRINCIPAL MANUSCRIPT SOURCES

According to Sørensen, 'Detailed and carefully composed diary notes are the basis for Nansen's books on travel from the first trip in the *Viking* in 1882 to the last through Armenia and the Caucasus in 1925.' Sørensen quoted extensively from these diaries, especially in regard to Nansen's temperament and feelings.

Fridtjof Nansen's diaries and other papers are held in the Department of Manuscripts in the Library of the University of Oslo.

C.I. Jackson lives in Hamden, Connecticut, and is a director of Chreod Ltd, Ottawa, Ontario.

Chronology

1861   Born on his mother's estate, Store Frøen, near Oslo (then Christiania), 10 October

1877   Mother died

1880   Entered University of Christiania, specialising in zoology

1882   First Arctic cruise to seas around Jan Mayen island and east Greenland in sealer *Viking* (March–July)

1883	Appointed curator of zoological collection, Bergen Museum
1885	Father died; Nansen took year's sabbatical from Bergen to study in Italy, first at Padua and then in Naples (Stazione Zoologica)
1888	Sailed from Christiania (2 May) to lead Greenland Expedition, Reached east Greenland pack ice 5 June. Disembarked from *Jason* into small boats at 65° N (17 July). Carried southwards in pack ice to 62° before reaching land (29 July). Reached Umivik area (64°) 10 August. Embarked on crossing of ice sheet 14 August. Reached ice-free land in west Greenland 24 September. Nansen and Sverdrup reached Godthåb 3 October
1889	Expedition sailed from Godthåb in *Hvidbjörnen* about 15 April; reached Copenhagen 21 May and Christiania 30 May. Nansen married Eva Helene Sars. European lecture tour. Appointed Curator of Zoology at Christiania University
1893	Daughter Liv born 8 January (d. 1959); *Fram* sailed from Christiania in June and became frozen in Arctic pack ice on 22 September
1895	Nansen and Johansen left *Fram* on 14 March, reaching then 'farthest north' at 86°13.6′N on 8 April. Entered winter quarters on 'Frederick Jackson Island', Franz Josef Land (O. Dzheksona, Zemlya Frantsa-Iosifa) on 26 August
1896	Nansen and Johansen left winter quarters 19 May; met Jackson expedition members 17 June. Sailed from Cape Flora 7 August. *Fram* released from ice 13 August. Nansen and Johansen rejoined *Fram* at Tromsø. Expedition returned to Christiania 9 September
1897	Lecture tour in North America. Son Kåre born (d. 1964)
1900	Daughter Irmelin born
1901	Son Odd born (d. 1973)
1903	Son Åsmund born (d. 1913)
1906–08	Norwegian Minister in London. Negotiated the Integrity Treaty (signed Christiania 2 November 1907)
1907	Wife Eva died
1917–18	In Washington, DC, to negotiate trade agreement that ensured Norwegian food imports during wartime
1919	Married Sigrun Munthe (1869–1957), neé Sandberg, previously the wife of the Norwegian painter Gerhard Munthe (divorced 1918)
1920	Accepted appointment (initially on short-term basis) as League of Nations Commissioner for Refugees. Continued in this position, and as Norwegian delegate to the League, throughout most of the 1920s
1922	Awarded Nobel Peace Prize
1930	Died at Oslo home, 13 May

Antonio Raimondi del Acqua

1826–1890

Leon I. Yacher

1. Education, Life and Work

Antonio Raimondi del Acqua was born in Milan, Italy, on 20 December 1826. His father, Enrique Raimondi, and his mother, Rebeca Dell'Acqua, were the parents of seven children; Antonio was the penultimate. He and his younger brother Timoleon alone reached adulthood; Timoleon lived four years longer than Antonio.

During his childhood, Antonio Raimondi showed keen interest in the natural sciences. At the age of 13, he purchased the works of Georges Buffon, the great French naturalist. It is possible that he was initially influenced by Buffon's work. Very early in his life, Raimondi travelled the hinterland of Milan. Later he travelled extensively in Europe, while reading the works of numerous scientists and travellers. He spent many hours reading the works of James Cook, Louis Antoine de Bougainville, Alexander Von Humboldt and Dumont d'Urville. He frequented zoological parks and museums of natural history.

Little is known about his family life. An older brother studied diplomacy in Vienna, but upon starting his diplomatic career drowned while bathing in the Gulf of Spezia. Timoleon joined the church; there he received a formal schooling and subsequently undertook international travel as church emissary first to Oceania and then for thirty-six years in Hong Kong. During a visit to Rome, Timoleon became a bishop.

When Raimondi left Italy on 8 January 1850, at the age of 24, he left behind a land experiencing bloodshed. Europe was undergoing a wave of revolution; absolutism was in vogue, and the Roman Empire, proclaimed in 1849, brought instability and fear to a people in the process of unification. He participated as a soldier in the revolution of Milan and the Lombardy campaign against the Austrians. With Austria firmly in control, and a divided homeland, Raimondi went to Genoa, where he started his voyage to Peru, crossing the Atlantic in the SS *Industria*. The seven-month-long trip took him around Cape Horn, where he witnessed a dreadful storm. He chose Peru because 'the richness and varied territory would provide an area of exploration and study'. The land of Peru seemed to provide a variety of conditions which intrigued him. Soon after arrival he wrote:

> in the sands of the coast, I can find the arid deserts of Africa; in the vastness of the Puna, the monotonous steppes of Asia; in the high peaks of the Cordillera [of the Andes], the frigid polar regions; and, in the thick rainforest of the Montana, the active and lush tropic vegetation; these decided for me to travel to Peru as my field of exploration and study. (Bustos Chavez, p. 17)

Doubtless several factors played a role in his decision to migrate. Raimondi was aware of the literature and of the presence of his contemporaries. He knew that several of the Latin American countries were being investigated and settled by Europeans while Peru remained less known. He was aware of those who had contributed to a knowledge of Peru, including Father Feuillée, Ulloa, H. Ruiz Lopez, J. Pavon and J. Dombey. Each of these individuals indirectly contributed to his decision to select Peru as his arena for study.

Upon arrival in Callao, Lima's port city, on 28 July 1850, he was met by José Cayetano Heredia, an influential scientist from Piura (northern Peru), and director of the College of Independence, later transformed into the School of Medicine. Heredia provided food, shelter and employment for Raimondi and was otherwise influential as mentor. It is no coincidence that Raimondi's first work concerned Ancash, a 'department' located north of Lima, *en route* to Piura. Raimondi assumed residence in Huaraz, while traversing Ancash numerous times. In Huaraz he met and married Adela Loli on 1 September 1869; they had three children, Maria, Enrique and Elvira. A few years later, Raimondi became the first to teach analytical chemistry in the College of Independence, a position created by Cayetano Heredia. Nearing death, Heredia gave his savings to Raimondi to enable him to introduce analytical chemistry to the University of San Marcos in Lima.

During the forty years that he travelled throughout Peru, Raimondi catalogued more than sixty species of flora, many unknown until then, while collecting plants, minerals, fossils and (dissected) animals. Furthermore, he wrote scrupulously about what he noted in his travels. Diaries, articles, books, maps and plans are among the many items he published. In Milan he had been trained in geography, geology, mineralogy, zoology, botany, palae-ontology, meteorology, hydrography, topography, seismology, archaeology, anthropology, philology and history.

During his first seven years in Peru he took a number of short trips, exploring the Lima surrounds, and more especially the desert region. In 1885 he took to higher elevations, visiting Chanchamayo and Vitoc. He journeyed to Tingo Maria in 1857. His first preliminary major trip started on 26 March 1859, when he travelled from Trujillo to Cajamarca, Chachapoyas and Moyobamba. By canoe he travelled the Huallaga, Marañón and Ucayali rivers. He returned to Trujillo by way of Moyobamba. After crossing Ancash and Huánuco, he returned to Lima in February 1861, twenty-three months after departure. Realizing that he had missed numerous places, he planned three more extended trips: one through the southern part of Peru, followed by one through the centre of the country and, finally, a major sortie across the northern region.

The first major trip commenced in May 1862, and finished on 31 January 1866, a three-and-a-half-year odyssey. He 'travelled this long trip in part on beast and in part on foot'. The second trip began in August 1866, when he headed to the confluence of the Mantaro and Apurimac rivers, and ended in December of that year. The final trip commenced in June 1867, and lasted two years; he returned after a 1,300-mile trip that ended in June 1869. So he lived for nineteen years, experiencing dangers of all types, while amassing notes, and readying *El Perú*, the most complete work on Peru at the time. Upon finishing his first volume of *El Perú*, he wrote:

> Here I am, after an infinite number of dangers, returning healthy in body and spirit to this post of salvation. Here am I at last, rich in materials and knowledge about the country, a treasure for me of incalculable value, picked up, I can say, at the expense of my blood and my youth. Actually only one thought torments me, and it is the continual doubt that I may not live long enough to complete my undertaking. ('Significativos', p. 19)

Commissioned by the Peruvian government, he was to travel (largely by mule) throughout the country and provide an accurate account of all resources, both cultural and physical. Raimondi quickly learned that he needed a variety of maps to help with this project. Not surprisingly, he found none that he could use. As a result he set out to map Peru topographically. Under difficult circumstances, he accomplished his assigned task.

In addition to the publication of the aforementioned five volumes of *El Perú*, significant socio-demographic and physical characteristics were recorded throughout the country. He also published regional studies, most of which remain little known to geographers and others (for example, *La provincia litoral de Loreto* and *El departamento de Ancash y sus riquezas minerales*).

Among Peruvian scientists Raimondi's name is well known. Schools and streets have been called after him. In Lima, a major private high school carries his name; a province was called after him in the Department of Ancash; and a district carries his name within the Province of Coronel Portillo in the present-day Department of Ucayali. The Urubamba and Tambo rivers, which form the Ucayali River (eventually joining the Marañón to form the Amazon River), traverse the Raimondi district. At Iquitos, in the heart of the Peruvian Amazon, not far from the birthplace of the Amazon River, his name is fixed throughout the landscape. Raimondi travelled through these areas and left his imprint.

In 1871, the Italian Geographical Society awarded Raimondi its Gold Medal. The Italian Anthropological Society made him an honorary member, and he was made corresponding member of the geographical societies of London, New York, Lisbon and Madrid, the Anthropological Society of London, and the Humboldt Society of Mexico (Janni, p. 283). A few well-known scientists recognised Raimondi's stature. Elisée Reclus, for example, said that Raimondi was for Peru 'almost a revelation' (Riccardi, 1947, p. 23). In spite of this, beyond Peru, little is known of him or his accomplishments.

At present Raimondi's books are used not only as reference, but perhaps more importantly as historical documents. His accounts of what he saw were so meticulous that present-day scholars adopt his work as a reliable source.

Alvarez-Brun, in the preface to his book about Ancash, explained that at the time of its 1965 publication, Raimondi's book, published in 1873, was one of only two dealing with Ancash as a unit. Several lesser studies existed, but Raimondi's work remained the only major contribution concerning this area until 1962, when the second work appeared – a mimeographed report, which is no longer available. Raimondi's book on Ancash was essentially a geo-morphological study, although matters of a political-administrative nature were included.

Raimondi was equally prolific as a cartographer. His map of Peru, published in 1877, served as the base for the national map portraying political, industrial and land-use information on the country. It continues to serve as a source of information to this day.

As a collector of specimens, he left behind 2,994 mineral samples, 2,000 fossil samples, 700 rock samples, over 20,000 plant samples, 1,265 bird specimens, 4,000 insect specimens, and 102 skulls of ancient Peruvians. And he left behind drawings of archaeological remains from both the Incan and pre-Incan periods ('Significativos', p. 21).

He died in San Pedro de Lloc, in the Department of La Libertad, northern Peru, on 26 October 1890, at 64 years of age, never having returned to Italy. Raimondi's remains can be found in the Mausoleum of Lima.

2. Scientific Ideas and Geographical Thought

Many in Peru refer to Raimondi as a genius. His keen awareness of important geographic concepts led him to observe such things as transportation development and the potential contribution of this to the development of a region. He observed that building a rail network between Chimbote and Huaraz would permit mineral exploitation, facilitating economic growth. This suggestion was intended also to reduce costs, since at that time another rail line existed, though the latter route was slower and more expensive (the slope of descent and ascent on the mountain slope was great, and the railmen would ride the brakes on the way down and use a greater amount of combustible fuel on the way up).

In 1857 he published his two-volume *Elementos de botánica*, in which many hitherto uncatalogued plants were introduced. Eventually he published many books, notes, maps and diagrams, revealing the detail of his labour. His versatility led him to catalogue numerous minerals previously unknown to the scientific community of Peru. As a result of his contributions, Raimondi did not go unnoticed by officials of the Peruvian government. He was contracted to survey and map the country, a task as then not accomplished. His method provided Peru with a much-needed, rigorous scientific approach. Raimondi proceeded to travel and survey the country and then started to write the five-volume *El Perú* (he originally planned to publish twenty volumes). This was virtually a geographic handbook of the country, and contributed to the knowledge of Peru. Included in the work were data concerning land forms, climates, soils, vegetation, population, and economic and transportation development. Raimondi's mode of transportation was crude and difficult: he walked or rode by mule. It should be noted that because of rudimentary conditions, he was able to record information in a detailed manner that might be bypassed by newer technological methods.

Raimondi recognized the significant role of maps and thus designed them carefully, portraying detailed geographic patterns. Although many of his maps depicted land-form types, he also used thematic maps to complement his essays. The land-form maps were largely qualitative; his concern was with the shape of the land form. Many of his maps also included quantitative information. The symbols he included in his mapping were simple and descriptive.

Raimondi's ideas in geography were diverse. Trained in such a variety of fields of learning, he brought to his work a range of approaches to specific problems. He was passionate about studying every aspect of Peru's physical and human landscape, and was particularly interested in advancing the understanding of the country's geography. He remained abreast of the changes taking place not only within Peru, but also in the geographic literature. Though he did not live to see his work finished, he was a figure who inspired others to follow where he led.

3. Influence and Spread of Ideas

Raimondi brought with him from Italy a body of knowledge hardly known in Peru. His teaching of various subjects contributed much to a deprived population. Furthermore, his findings provided a knowledge of what had been previously unknown in the country. The Peruvian government, recognizing the value of Raimondi's work, financed some of his travels.

Raimondi encouraged the study of science in Peru. Geographers and other scientists continue to use much of his work, especially in remote locations where even now little is known. His dedication to Peru is evident from his refusal to send the findings of his labour to Italy. After numerous requests from that country, Raimondi responded by saying of his work that it 'belong[ed] to Peru, it should remain and have Peru's luck' (Janni, p. 281).

Many of his contemporaries honoured Raimondi and his accomplishments: Breythaupt dedicated a mineral to him, Lisson a fossil, and Taczanowski a bird. Others, like the German botanist Weberbauer, dedicated a book to Raimondi. Just prior to Raimondi's death the Peruvian government voted to continue funding him in order that he might finish his work. The monies allowed Raimondi to maintain two institutions in which his labatorary and collections could be housed. At this point he started work on both the 34-sheet map of Peru project at a scale of 1:500,000 (none previously existed), and the fourth volume of *El Peru* (published posthumously).

Raimondi helped shape the field of geography in Peru. His ideas remain a strong force in the country's schools and universities. Today, while the country grapples with instability, and the geographic community faces difficult times, Raimondi's work is still considered monumental. Nothing else exists that can compare to his attainments.

Raimondi's library in Lima houses a large collection of materials that commemorate his accomplishments, and the Geographical Society of Lima contains a large number of his original materials and maps. Had he attempted to disseminate his work beyond Peru, he would have become better known.

Bibliography and Sources

1. REFERENCES ON ANTONIO RAIMONDI

Balta, José, *La labor de Raimondi*, Lima, Torres, 1926.

Bustos Chavez, Cristobal, *La vida y la obra del sabio Antonio Raimondi Dell'Acqua*, Lima, 1949.

Gerbi, Antonello, 'Romanticismo e peruanesimo di Antonio Raimondi', in *Le Americhe, Storie di viaggiatori italiani*, Milan, Electa, 1988, 162–7.

Janni, Ettore, *Vita de Antonio Raimondi*, Milano, Ed. Alpes, 1940 (translated into Spanish four years later).

Llona, Emiliano, *La obra de Raymondi*, Lima, Bacigalupi, 1884.

Patrón, Pablo, *Observaciones sobre la obra 'El Perú' del señor Antonio Raimondi*, Lima, Rosay, 1902.

Porras Barrenechea, Raúl, 'El siglo XIX: Bolognesi, Garibaldi y Raimondi', in *Los Viajeros italianos en el Perú*, Lima, Ecos, 1957, 76–90.

Raimondi's Library, archival holdings, Lima, Peru.

Riccardi, Riccardo, 'Antonio Raimondi, geógrafo del Perú (1826–1890)', in *Le vie d'Italia e d'America Latina*, 1927, 103–8.

Riccardi, Riccardo, 'Antonio Raimondi, esploratore e geógrafo del Perú', in *Bollettino della Societá Geografica Italiana*, Vol. 12 (1947), 9–23.

'Significativos homenajes tributados a la memoria de Antonio Raimondi', *Boletín de la Sociedad Geográfica de Lima*, Vol. 72 (1955), 8–31.

Sociedad Geográfica de Lima, archival holdings, Lima, Peru.

2. SELECT BIBLIOGRAPHY

1857   *Elementos de Botánica* [*Elements of Botany*], 2 vols, 600 pp.

1862   *La provincia litoral de Loreto* [*The Litoral Province of Loreto*], Iquitos, Imprenta El Oriente, 159 pp. Reprinted 1942.

1864   *Análisis de las aguas termales de Yura, aguas minerales de Jesús y aguas potables de Arequipa* [*Analysis of the Yura Thermal Waters, the Mineral Waters of Jesus and the Waters of Arequipa*], Arequipa, Imprenta de Francisco Ibanez.

1873   *El departamento de Ancash y sus riquezas minerales* [*The Department of Ancash and Its Rich Mineral Resources*], Lima, Imprenta El Nacional, 651 p.

1874   *El Perú*, Lima, Imprenta del Estado, Vol. 1, 444 pp.

1875   'Observaciones al dictamen de los señores Disneros y Garcia en la cuestión relativa al Salitre', Lima, La Opinión Nacional, 16 pp.

1876   *El Perú*, Lima, Imprenta del Estado, Vol. 2, 475 pp.

1878   *Minerales del Perú*, Vol. 1, 305 pp.

1879   *El Perú*, Lima, Imprenta del Estado, Vol. 3, 714 pp.

1882   *Mineral Waters of Peru*, Lima, 210 pp.

1883   *Minas de oro de Carabaya, Puno* [*Gold Mine of Carabaya, Puno*], ed. Carlos Paz Soldan, Lima, Anales de C.C. y de Minas, 32 pp.

1884   *Potable Waters of Peru*, Lima, Anales de C.C. y de Minas, 155 pp.

1885   *Memoria sobre Cerro de Pasco y la montaña de Chanchamayo* [*Account of Cerro de Pasco and the Chanchamayo Mountain*], Lima, Imprenta de la Merced, 39 pp.; tables and maps.

1902   'Apuntes sobre el mineral de Hualgayoc', in *Perú: Estudios Mineralogicos-geológicos* [*Peru: Minerological-geological Studies*], Lima, Imprenta Gil, 506–13.

1902   'Islas, islotes y rocas del Perú', in *Perú: Estudios Mineralogicos y Geológicos* [*Peru: Minerological-geological Studies*], Lima, Imprenta Gil, 398–432.

Not included are numerous newspaper articles, notes and maps.

Leon I. Yacher, born and raised in Peru, is Associate Professor of Geography at Southern Connecticut State University, New Haven, Connecticut 06515-1330, USA.

Chronology

1826   Born in Milan, Italy, 20 December

1850   Departed Italy on the SS *Industria*, 8 January

1850   Arrived in Callao, Lima's port city, 28 July

1852   Crossed the Andes for the first time, visiting Vitoc and Chanchamayo

1855   Left for Tingo Maria, Huallaga and Huánuco

1857   Ended the trip and research project

1857   *Elementos de Botánica* published

1859   First preliminary major trip started on 26 March

1861   Returned to Lima in February

1862   First major trip started during May

1866   First trip finished on 31 January; second trip began in August, ended in December

1867   Final major trip commenced in June

1869   Returned in June

1869   Married Adela Loli on 1 September

1871   Gold Medal, Italian Geographical Society, accepted *in absentia*

1874   *El Perú*, Vol. 1, published

1876   *El Perú*, Vol. 2, published

1877   Map of Peru published

1878   *Minerales del Perú*, Vol. 1, published

1879   *El Perú*, Vol. 3, published

1883   First of 34 sheets of larger-scale maps of Peru published

1890   Died in San Pedro de Lloc, 26 October

1902   *El Perú*, Vol. 4, published posthumously

1913   *El Perú*, Vol. 5, published posthumously

Two French Geographers
Paul Reclus and Louis Cuisinier

Gary S. Dunbar and Louise Rapacka

It is indeed fitting to have brief sketches of Paul Reclus and Louis Cuisinier in *Geographers: Biobibliographical Studies*, because they were nephew and grandson, respectively, of the great French geographer Elisée Reclus (1830–1905), whose life was previously described in this publication (Volume 3, 1979). Another member of the *tribu*, Franz Schrader (1844–1924), Elisée's cousin, has also been covered in *Geographers* (Volume 1, 1977). Paul Reclus and Louis Cuisinier deserve notice in these pages, because much of their work was of geographical interest. Louise Rapacka is the daughter of Louis Cuisinier, and she knew 'Uncle' Paul Reclus well. Another geographer in the Reclus family, Elisée's brother Onésime (1837–1916), would also be a deserving subject for biographical treatment.

Paul Reclus

1858–1941

1. Education, Life and Work

Paul Reclus was born in the Paris suburb of Neuilly-sur-Seine on 25 May 1858, the son of Elie and Noémi Reclus. Elie had returned to France in 1855 and married his cousin Noémi the following year. Elie's younger brother Elisée, who was to become a famous geographer, returned to France in 1857, married in 1858, and thenceforth tried to stay as close to Elie as possible; indeed, the brothers and their families shared the same apartment throughout the 1860s, down to the suppression of the Paris Commune in the spring of 1871. Elie briefly served as director of the Bibliothèque Nationale during the Commune but avoided capture by the French army and so escaped to Switzerland. Elisée, as a foot-soldier in the Commune's ragtag army, was captured, imprisoned for fourteen months, and then exiled to Switzerland. Although it was Elie's presence in Switzerland that guided Elisée's choice, the brothers did not live so closely there; Elie had settled in Zurich, whose Germanic character did not suit Elisée, and so he chose to live elsewhere, first in Lugano near the Italian border and later in Vevey and Clarens at the eastern end of Lake Geneva. In an autobiographical essay in 1938, Paul claimed that his father chose Zurich because he recognized that his 13-year-old son's talents seemed to be more scientific than literary and that Zurich would offer superior educational opportunities for him; otherwise, the family might have gone to Constantinople.

Paul Reclus was thus able to grow up in an intellectually rich, if materially poor, household, surrounded by a large circle of family members and their friends. In the 1860s the Reclus brothers had maintained a weekly salon in their Paris apartment, according to the fashion of the time, but theirs was an unusually eclectic one, probably embracing a greater number of political leftists than was the case with the more artistic and literary salons of the city. The visitors were mostly Parisian socialists, but there were others, even a Basuto prince, 'singing to us a lyric of his country'. Becoming increasingly disillusioned with the slow progress toward true democratization and liberalization during the Second Empire, Elie and Elisée drifted leftward during the 1860s and became caught in the embrace of the charismatic Russian anarchist Mikhail Bakunin, whose principles they espoused for the rest of their lives, Elisée more enthusiastically than his mild-mannered and pessimistic brother. Not surprisingly, Paul absorbed the family characteristics and grew up to be very much like his father, even to the extent of inheriting Elie's only 'fault', which was his excessive modesty.

Paul Reclus spent his adolescent years in Zurich and then returned to Paris for his higher education (1877–80) at the Ecole Centrale, the French *'grande école'* for the training of industrial engineers and managers. After a year of military service, he worked at various engineering jobs, married in 1885, and got involved with anarchist activities in Paris. The police tended to track the anarchists rather loosely but started to crack down after the Vaillant Affair on 9 December 1893, when an anarchist threw a bomb into the Chamber of Deputies. As 'elder statesmen' of the anarchist cause, Elie and Elisée Reclus were not immune from police harassment, and they voluntarily exiled themselves to Brussels for the remaining decade of their lives. As a younger and therefore potentially more dangerous anarchist, Paul would have been imprisoned, but he escaped to Great Britain, where he lived for nine years under the assumed name of Georges Guyou. He went first to London to work as a book-binder for the Cobden-Sandersons, friends of Elisée Reclus and his Russian colleague Peter Kropotkin, a fellow geographer and anarchist. On Kropotkin's recommendation, Paul went to Edinburgh in 1896 to work for Patrick Geddes on various cartographic projects in connection with Geddes's new geographical museum, the Outlook Tower. (See the biographical sketch of Geddes in *Geographers*, Vol. 2, 1978.) Apart from making maps and relief models for the Tower, Paul was engaged in co-authoring papers on cartographic subjects with his uncle Elisée and especially in drafting elaborate plans for the latter's Great Globe. Although never completed, the Globe, originally planned (at a scale anywhere from 1:50,000 to 1:1,000,000) for the Paris Exposition of 1900, was an obsessive interest of the Reclus until Elisée's death in 1905, and even longer for Patrick Geddes. (See especially G.S. Dunbar, 'Elisée Reclus and the Great Globe', *Scottish Geographical Magazine*, Vol. 90 (1974), 57–66.)

Paul Reclus taught geography, science and French in the high school in Peebles, Scotland, from 1901 to 1903, when he was called to Brussels to work in the Geographical Institute that Elisée had founded in 1898 in the New University, an institution that was created in 1894 by the Reclus brothers and some like-minded Belgian colleagues, mostly leftists of various hues. The New University was not allowed to give degrees, and its institutes had to be self-supporting. No one was paid for teaching in the Geographical Institute, and it suffered a threadbare existence until it was finally closed at the outbreak of war in 1914. The Geographical Institute was supported largely by Elisée Reclus through his royalties from publications and from new cartographic projects, including the illustration of his posthumous six-volume work, *L'Homme et la terre*. Paul Reclus taught English, German and geography at the New University. He was barely reunited with his father in Brussels when the latter died in February 1904, and his mother died in July 1905, surviving her brother-in-law (cousin) Elisée by little more than a week. As in life, the brothers are united in death, buried in the same plot apart from their wives. After Elie's death, Elisée declined rapidly in health, but he had full faith that Paul could take over his work. As he said to his friend Nadar, the famed photographer, 'My nephew . . . will complete me, enlarge me, and extend me.' Paul lived up to Elisée's expectations by completing all his writing projects and by taking over the direction of the Geographical Institute.

One of Elisée Reclus' last publications was the *Introduction* to the seven-volume *Dictionnaire géographique et administratif de la France*, edited by Paul Joanne. The introductory volume (1905) was actually the last volume published in the series, which began in 1889. It was similar to the earlier geographical introductions that Elisée and Elie had done for Paul Joanne's father, Adolphe, in the 1860s. Although Paul Reclus' name does not appear in the book, he actually composed about three-quarters of the work.

Another project that was completed by Paul but bears only his uncle's name was *Les volcans de la terre*, which was published in three parts from 1906 to 1910 by the Belgian Society of Astronomy, Meteorology and Geophysics. Elisée's lifelong interest in volcanoes was rekindled by the eruption of Mont Pelée in Martinique in 1902, and in 1903 he proposed an 'Atlas of Volcanoes', but the final product fell far short of his grandiose scheme. Only southwest Asia and southern Europe were covered, with emphasis on Italy, and the work was almost purely factual and descriptive.

The largest project that Elisée left for Paul to complete was the panoramic social and historical geography, *L'Homme et la terre*, that was published in six volumes in the three years following Elisée's death. In this work, which consisted of 4,500 manuscript pages by the time of its completion in 1904, Elisée used his lectures on comparative geography in the New University and also his latest journal articles. The first fascicule, or weekly instalment, was published on 15 April 1905, and Elisée died less than three months later, leaving Paul to complete the massive work. Elisée had begged Paul to rework the manuscript and to stamp his own personality on it, but, true to form, the nephew tried to follow his uncle's plan as closely as possible and to avoid taking any credit for himself. 'It may take me days for a single page', he told his friend Patrick Geddes in 1906, 'but I don't know of anyone who knew better what Elisée's aim was, and how far history, geography, sociology and literature can be mixed.'

Paul took over the administration of the Geographical Institute at the New University after Elisée's death and continued to teach a lecture course on economic geography. The staff of the Institute dwindled after 1905, and its monograph series ended with Edgar Sacré's 15-page paper on Esperanto (a subject of great interest to anarchists) in that year. Paul also taught in the Lycée français of Brussels from 1908 to 1913. He and his family fled Brussels in October 1914, and the Institute closed abruptly at that time, never to be revived. The New University was expected to resume operations after the war, but in fact only its Institut des hautes études reopened – and continues to this day. Paul removed his family to France, and he spent most of the war working in a munitions factory in Sevran, a northeastern suburb of Paris. In 1919 they moved into a house left to Paul by his aunt Noémi Mangé in Domme, a small village in southwestern France. Paul remodelled an old windmill in Domme to serve as a regional museum much like Patrick Geddes's Outlook Tower in Edinburgh. From 1914 to 1920 a Japanese anarchist, Ishikawa Sanshirō, lived with the Reclus family. Ishikawa made plans for an Elisée Reclus Geographical Institute to be established in Japan and arranged for the purchase of the Brussels Institute library (some 60,000 books) in 1922, but the library was utterly consumed in the fire accompanying the great Tokyo earthquake of 1 September 1923, and the institute plans were scrapped, although Ishikawa (1876–1956) continued to write and lecture about Elisée Reclus in Japan and in 1948 published a Reclus biography. (See G.S. Dunbar, 'Elisée Reclus' Japanese disciple, Ishikawa Sanshirō', *History of Geography Newsletter*, No. 2 (1982), 18–20.)

In the 1920s Paul Reclus worked on several projects that would keep his uncle's name alive in France and in the world generally. After the death of his aunt Louise, who had been Elisée's secretary and editor of his *Correspondance*, Paul edited the third and final volume, which was published in 1925. He gave enormous help to Elisée's biographers, the anarchist historian Max Nettlau and the libertarian printer Joseph Ishill. Nettlau published *Elisée Reclus, Anarchist und Gelehrter* in 1928, and the enlarged Spanish edition published the following year contains interesting data supplied by Paul Reclus too late for the German edition. Ishill printed a number of essays by Paul and others under the title *Elisée and Elie Reclus: In Memoriam* in 1927. Paul continued to gather materials on his father and uncle. He completed a book-length manuscript by the outbreak of the Second World War, but it was not published until 1964, when it was issued by Les Amis d'Elisée Reclus (Paul's sons Jacques and Michel and their associates) under the revealing title, *Les frères Elie et Elisée Reclus, ou, du protestantisme à l'anarchisme*.

After the publication of the Ishill book, Paul went to work on a revised edition of Elisée's *L'Homme et la terre*, which Paul had seen through the press after his uncle's death. With the help of a young geographer, Aimé Perpillou, who made a number of small changes in the first thirty-eight chapters, Paul reduced the work by about one fifth and deleted numerous illustrations, but added about twenty new maps and rewrote the final twelve chapters to cover the events of the last quarter century. The new edition was published by Albin Michel in 1930–31 in three volumes.

In 1924 Patrick Geddes made plans for the Collège des Ecossais, a residence hall for foreign students at the University of Montpellier in southern France.

Paul Reclus was invited to join the enterprise as 'director of studies'. After Geddes's death in 1932, Paul and T.R. Marr managed to hold the college together after a fashion, and it limped along until the advent of war in 1939.

Throughout his adult life Paul was dedicated to the production and dissemination of anarchist propaganda. For a full half century he contributed to the Parisian journal *La révolte* and its successors *Les temps nouveaux* and *Plus loin*. Paul died on 19 January 1941, but his sons and other members of the Reclus family have kept his spirit alive, and through them the wider world has been reminded of his accomplishments and especially those of his better-known uncle Elisée.

2. Scientific Ideas and Geographical Thought

Paul Reclus' geographical writings, even more than those of his uncle Elisée, were rarely speculative or theoretical. His practical engineering training, his 'pathological modesty' (in the words of his son Jacques), and his strict adherence to following what he interpreted to be Elisée's line of thought tended to curb any notions of deviation or innovation that he might have harboured. Although he was a man of great intellectual ability, he always subordinated his own ideas to those of his father and uncle and his friend Patrick Geddes. Perhaps the clearest statement of Paul's conception of geography was in an autobiographical essay that appeared in *Plus loin* in 1938. Here he asserted that his basic programme of geography – to know the earth and to understand how humans can mould it for their use and organize themselves in order to live there intelligently – involved a synthesis, not only of the kinds of geography that one is traditionally taught, but also of engineering studies, the regional museum and the anarchist mentality.

Privately, Paul did not always agree with Elisée's ideas, but he never exposed their differences in print. In 1928 Paul published a stout defence of his uncle's geographical works, even while admitting their obsolescence. It had been six decades since Elisée published his first great compendium of physical geography, *La terre*, but the overarching themes retained their significance, even though the facts had been superseded. Paul turned his uncle's lack of fieldwork and original observation into a virtue by claiming that his writings thereby possessed greater sweep and homogeneity. By remaining a generalist, Elisée achieved a grander synthesis and thereby occupied a higher plane than the narrow specialists who succeeded him. 'He was a geographer whom everyone can read with pleasure and without special education . . . He did not play too much on only a single string of his lyre' ('Valeur actuelle de la science des frères Reclus', p. 3).

3. Influence and Spread of Ideas

Several factors conspired to limit Paul Reclus' reach or fame: his preference for remaining in the shadow of others; his association with unconventional institutions (the Outlook Tower, the New University and the Scots College) and his consequent lack of students or protégés; his slight literary output; and his aloofness from engaging in the building of the emerging discipline of geography. His greatest contribution was in completing Elisée Reclus' works and thereby further enhancing his uncle's reputation. Elisée's accomplishments were truly remarkable, but they owed much to the support of his devoted family, particularly his brothers Elie and Onésime, his sister Louise, and, above all, his nephew Paul. Elisée's words to his friend Nadar in 1904 were truly prophetic: 'My nephew . . . will complete me, enlarge me, and extend me.' In no small measure, Paul Reclus contributed to the enduring reputation that his uncle has enjoyed throughout the twentieth century.

Bibliography and Sources

1. MANUSCRIPT MATERIALS

Unpublished letters to or from Paul Reclus and other manuscript materials relating to him can be found scattered through various repositories, such as the Patrick Geddes Papers in the National Library of Scotland, Edinburgh; Autographes Félix et Paul Nadar in the Bibliothèque Nationale, Paris; Correspondance et papiers d'Elisée Reclus in the Bibliothèque Nationale; Fonds Elisée Reclus in the Institut français d'histoire sociale, Paris; the Nettlau Archives in the International Institute of Social History, Amsterdam; and the Joseph Ishill Collection at Harvard University.

2. WORKS ABOUT PAUL RECLUS

1894   *Album photographique des individus qui doivent être l'objet d'une surveillance spéciale aux frontières*, Paris, August 1894. Photograph of Paul Reclus on p. 10 and description on p. 22 ('Height 1 m 67 [*c.* 5′6″], black hair and beard, large nose, pale complexion, slight build, brown eyes, energetic manner, often wearing a black hat': Archives Préfecture de Police, Paris). Mme Rapacka says (1993) that the police description of Paul Reclus was incorrect in at least two particulars: his eyes were blue, and his height was underestimated by about seven centimetres.

1938   'Synthèse d'un individu' [Autobiographical essay], *Plus loin*, No. 156 (April), 1–5.

1964   'Paul Reclus: notice biographique', in *Les frères Elie et Elisée Reclus, ou, du protestantisme à l'anarchisme*, Paris, Les Amis d'Elisée Reclus, 203. Photograph of Paul Reclus in 1906 facing p. 194. Book essentially completed by Paul Reclus in 1939, and published (with additions such as the biographical note on him) by his sons Michel and Jacques in 1964.

1977   Maitron, Jean (ed.), *Dictionnaire biographique du mouvement ouvrier français*, 3rd part, *1871–1914, De la Commune à la Grande Guerre*, Vol. 15, *Ras–Z*, Paris, Les éditions ouvrières, 1977, 19–20.

3. WORKS BY PAUL RECLUS

a. Works Ascribed to Elisée Reclus

After Paul joined his father and uncle Elisée in 1903, he helped the latter with his ongoing projects and finished them after Elisée died in 1905. Paul so thoroughly disguised his contributions that it is impossible to separate them from Elisée's writing.

1905   *Introduction*, introductory volume to Paul Joanne (ed.), *Dictionnaire géographique et administratif de la France*, 7 vols, Paris, Hachette, 1890–1905.

1905–08 *L'Homme et la terre*, 6 vols, Paris, Librairie Universelle. Only Elisée's name appeared on the title pages, but the volumes were largely reworked and finished by Paul Reclus. The first volume was published in October 1905, three months after Elisée's death. Second edition, abridged to three volumes and edited by G. Goujon, Aimé Perpillou and Paul Reclus, Paris, Albin Michel, Editeurs, 1930–31.

1906–10 *Les volcans de la terre*, 3 parts, Brussels, Société belge d'astronomie, de météorologie et du physique du globe.

1925   *Correspondance.* Paul Reclus edited the third and final volume of Elisée Reclus' correspondence, published in 1925 in Paris by Alfred Costes. The first two volumes were edited by Elisée's sister, Louise Dumesnil (1835–1917), and published in 1911 in Paris by Librairie Schleicher Frères.

b. Papers published by Paul Reclus under a Pseudonym ('Georges Guyou') with Elisée Reclus

1895   *Projet de construction d'un globe à l'échelle du cent-millième*, Paris, Edition de la Société Nouvelle, 16 pp. Pp. 3–9 by Elisée Reclus, pp. 10–16 by 'G.G.'.

1896–97 'D'un atlas à l'échelle uniforme', *Bulletin de la Société neuchâteloise de géographie*, Vol. 9, 159–64.

1900   'L'anarchie et l'église', *Les temps nouveaux*, No. 20. Reprinted as a pamphlet in 1901.

1901   'On a one-scaled atlas', *Bulletin of the American Bureau of Geography*, Vol. 2, No. 3 (September), 199–204. G.P. Reclus-Guyou (Paul Reclus) was listed as 'Associate Editor for Europe' of this short-lived journal.

c. Works Signed 'Georges Guyou' but Not Co-authored by Elisée Reclus

1898   *The Dreyfus Case*, Edinburgh, Patrick Geddes and Colleagues. 57 pp. Second revised edition (68 pp.) also 1898.

1902   *Un nouveau planétaire*, Université nouvelle, Institut géographique de Bruxelles, Publication No. 7, 13 pp.

d. Other Works by Paul Reclus

c. 1897   *The Hollow Globe: A New Geographical Apparatus*, Edinburgh, Patrick Geddes and Colleagues, The Outlook Tower, n.d., 4 pp.

1913   'Les progrès du français dans l'agglomération bruxelloise', *La géographie*, Vol. 28, No. 5 (15 November), 308–18

1922   *Nouvel atlas classique*, Brussels, Maison d'édition A. de Boeck. Apparently the atlas and accompanying text were largely the work of Paul Reclus, but his name does not appear. The work was ready for publication in 1915, but it was delayed by the war, and some twenty maps had to be redone before the atlas was finally issued in 1922.

1927   'A few recollections on the brothers Elie and Elisée Reclus', in Joseph Ishill (ed.), *Elisée and Elie Reclus: In Memoriam*, Berkeley Heights, NJ, The Oriole Press.

1928   'Valeur actuelle de la science des frères Reclus', *Le semeur*, Numéro spécial (8 February), 2–3.

Gary S. Dunbar is Professor of Geography (emeritus) at the University of California, Los Angeles. His address is 13 Church Street, Cooperstown, New York 13326.

Chronology

1858   Born at Neuilly-sur-Seine, 25 May

1871–77   Lived in Zurich, Switzerland

1877–80   Studied at Ecole Centrale des Arts et Manufactures in Paris and subsequently worked as an engineer

1885   Married Marguerite Wapler

1894   Fled to England

1896   Moved to Edinburgh to work with Patrick Geddes

1901–03   Taught at school in Peebles, Scotland

1903   Moved to Brussels to work with his uncle Elisée

1905   Death of Elisée Reclus, 4–5 July. Paul succeeded his uncle as director of the Geographical Institute of the New University of Brussels

1914   Returned to France to work as an engineer in the powder works at Sevran

1919   Moved to Domme, which remained his home base for the rest of his life

1927   Wife died. Subsequently he spent much time in Montpellier aiding Patrick Geddes with the Scots College

1932   Geddes died

1941   Paul Reclus died at Montpellier, 19 January

Louis Cuisinier

1883–1952

1. Education, Life and Work

The French earth scientist, mapmaker, and engineer Louis Cuisinier was born in Paris on 13 July 1883 and died there on 30 August 1952. He was the son of a chemist, Léon Cuisinier (1859–87), and Jeanne Reclus (1863–97), daughter of the geographer-anarchist Elisée Reclus. No civil or religious authority presided over the union of Louis' parents. This was an important event in the history of anarchism: it provided the opportunity for Elisée Reclus, who himself presided over the simultaneous marriage of his two daughters, to give a speech on *'l'union libre'*, later published as *Souvenir du 14 octobre 1882 . . .* (Paris, 1882). At the age of 4, Louis lost his father, and he was subsequently raised, with a three-year interruption, in his grandfather's home, first in Clarens, Switzerland, then in Nanterre, near Paris (1890–91), and then in Paris (1891–94), where Louis attended the Alsatian School. Having remarried, his mother went to live in Menton with her family, and Louis attended school there. When Louis' mother died on 21 March 1897, Elisée Reclus took her three children with him to Brussels and personally looked after their education, together with his sister Louise Dumesnil-Reclus. In Brussels Louis studied at the Van der Stock and Ramaeckers institutes, where mathematics was taught at a high level. He spent the academic year 1901–02 studying mathematics at the University of Liège, then moved to the University of Geneva, where he was a student from 1902 to 1907 in the Faculty of Sciences. There he came to the attention of Louis Duparc (1866–1932), Professor of Mineralogy, who warmly recommended his former student to mining companies.

Having spent his childhood and adolescence among eminent scientists, Louis Cuisinier acquired a passion for the earth sciences and for precise observation and scientific experimentation. He was educated according to

Elisée Reclus' principles, explained in *L'Homme et la terre*, Vol. 6, Chapter 11: the child should be brought up without the useless restrictions that are imposed by the conventions of the day; the teacher should be at the same time a father and a brother, capable of conveying enthusiasm for consistent work, physical exercise and harmony with nature; logical ways of thinking and expression should be acquired through drawing, geometry and algebra. As 'examinations are alien to learning' – performance alone matters – Louis had to sit for examinations only when he entered the universities of Liège and Geneva.

Apart from the missions linked with his work (see the list below under 'Bibliography and Sources'), and during which he wrote reports, drew maps and took photographs, Louis Cuisinier travelled alone, on foot and on bicycle, in Europe and throughout the world, continually making observations on the geology and physical and human geography of the regions traversed. Elisée Reclus cited one of Louis' observations in *L'Homme et la terre*, Vol. 1, p. 74, and photographs taken by Cuisinier in the Niger Valley and Russia were published in Vol. 6.

In addition to its renowned teaching, the University of Geneva gave Louis Cuisinier an insight into a world even more cosmopolitan than that of Brussels. Indeed, the university calendar for the winter term of 1906–07 ('Liste des autorités, professeurs, étudiants . . .') shows that, of the 1,634 students, only 405 were Swiss, whereas 809 came from the Russian Empire. That was a time when women, Jews and political dissidents could not gain entry to Russian universities. In Geneva one could mingle with the future revolutionaries of 1917, such as Lenin, Radek and Dzierzynski, as well as representatives from many other nations and political circles. Louis Cuisinier always remained profoundly influenced by the anarchism (impregnated with Protestantism) of his grandfather, and he held to its ethical requirements and its principles of freedom and individual responsibility. He did not, however, retain Elisée Reclus' political ideas, finding them inapplicable. He developed this position during his sojourn in Geneva, where he had the opportunity to compare diverse political theories.

An exceptional mountaineer, Louis Cuisinier was the first Frenchman to climb Mount Kilimanjaro in Tanzania. Without any companions, he reached the summit on 22 July 1925.

Louis Cuisinier spent the First World War with his Polish wife, Wanda Tutakiewicz, who had been a student in Geneva, in a civilian internment camp in Austria. He escaped from the camp in 1918 and returned to Switzerland on foot, avoiding the roads while crossing the Alps. He spent the Second World War in Indochina under very trying conditions. Thanks to some Laotians, he miraculously escaped the massacre of the French by the Japanese at the Phon Tiou tin mine in Laos. For nearly eight months (9 March to 1 November 1945) he lived in the Laotian bush.

A talented musician and a good writer, Louis Cuisinier was also an exceptional lecturer. In Geneva, his home base up to 1931, he gave lectures illustrated with slides. His last lecture, on Laos, took place on 19 April 1952 in the Natural History Museum in Paris. An excellent photographer, he made numerous stereo views on glass plates and prepared a catalogue of them. During the course of his fieldwork he designed and drew numerous maps and pen sketches.

Through the sponsorship of his relatives Paul Reclus and Joseph Kergomard, Louis Cuisinier became a member of the Paris Geographical Society. In fact, Paul Reclus played a very big role in the life of Louis Cuisinier, whose mother, Jeanne Reclus, Paul's first cousin, had married Paul's best friend and fellow student at the Ecole Centrale, Léon Cuisinier. When Louis was orphaned, his first tutor (Edouard Bouny) died in 1899 and was replaced in that function by Paul Reclus, who always displayed towards his ward a deep and faithful friendship, which ended only with Paul's death. Whenever Louis was in France, he spent at least one summer month at Paul's home in Domme, helping him in the building of a regional museum installed in an old windmill, inspired by Patrick Geddes's Outlook Tower in Edinburgh. Louis prepared maps and idealized geographical tableaux for the museum.

2. Scientific Ideas and Geographical Thought

Louis Cuisinier's writings display the breadth and depth of his scientific training and field expertise. They are largely descriptive accounts of the mining potential of tropical areas previously unknown or little known to scientific observers. His most important works were perhaps the topographic maps and geological diagrams that he made in Central Africa between Lake Tanganyika and the Congo River. These works remained secret because they were produced under contract to mining companies, and are unknown to geographers. His second greatest work was the geological study of the tin-mining area of the Nam Pha Thène river valley in central Laos from 1938 to 1946. A geological fault that he discovered in central Laos was given his name ('*faille Cuisinier*'). The duration of his sojourn enabled Cuisinier to elaborate a theory of the geological formation of that region, where he spent the Second World War in very difficult circumstances. Travelling on foot and on bicycle, he had the opportunity to study Laos in much greater detail than Jacques Fromaget did in his book and geological map on northern Indochina. Thus, at Ban Ca Tep, one finds not 'Triassic schists' but much granite, and the Uralo-Permian limestones are much more extensive than they appear on Fromaget's map. Cuisinier's work, however, has never been published.

3. Influence and Spread of Ideas

Like Paul Reclus, Louis Cuisinier lacked students to carry on his work, and his publications did not cause much of a stir among professional geographers, but both gain significance through their association with Elisée Reclus. Louis Cuisinier's life is interesting in its own right, but it gains added importance when it is fitted into the life history (prosopography, if you will) of the prolific Reclus family. Much of Cuisinier's work in the African and Asian tropics has retained its value simply because the political and economic conditions in those Third World countries have not permitted the luxury of original scientific investigation in the post-war era. Copies of Cuisinier's geological reports have been preserved in the archives of the Service de géologie in Vientiane, Laos. When Louis Cuisinier's grandson, Jan Rapacki, visited Laos in 1992, he was reassured of the value of his grandfather's work by the Laotians, since their knowledge of the geology of their own country has not progressed beyond what it was before the Second World War.

Bibliography and Sources

1. MANUSCRIPT MATERIALS

The Archives d'Outre-Mer in Aix-en-Provence possesses Louis Cuisinier's professional correspondence and field reports from 1910 to 1952, as well as manuscript maps, pen sketches and photographs (glass plates) illustrating his travels. In the Département des cartes et plans in the Bibliothèque Nationale, Paris, can be found his larger manuscript maps of the Congo Basin and Laos, along with some pen sketches. The monthly reports and associated 1:500,000 maps of his expedition to Manyéma, Belgian Congo, are in the Musée Royal de l'Afrique Centrale in Tervuren, Belgium.

The mission reports cover Louis Cuisinier's prospecting ventures in Indochina (1911–12, 1926–29, 1938–46 and 1949), Africa (French Guinea, 1907; Belgian Congo and Tanganyika, 1923–25; Ethiopia, 1930; Gabon and Congo, 1931; Ethiopia, 1933; Upper Volta, 1935; and Ivory Coast, 1936–37), Russia (1908 and 1914), Venezuela (1919 and 1920–21), and France (1939 and 1952).

2. WORKS BY LOUIS CUISINIER

1906   with Frédéric Reverdin, 'Nitration des dérivés *o*-acétylé et *o*-benzoylé des *p*-benzoyl et *p*-acétylaminophénols', *Archives des sciences physiques et naturelles*, Vol. 22, 449–57. Also published in *Bulletin de la Société chimique de Paris*, Vol. 35, 1255–60.

1914   'L'Indochine et le Tonkin', *Le globe*, Vol. 53, 11 pp.

1924   with Louis Duparc, 'Sur la géologie et les roches des environs de Bolivar (Vénézuéla)', *Livre jubilaire publié à l'occasion du cinquantenaire de la fondation de la Société géologique de Belgique*, Liège, 29 pp.

1926   'Un voyage au centre de l'Afrique', *La semaine littéraire*, Vol. 34, No. 1686, 202–3; No. 1690, 250–2; No. 1694, 296–9; No. 1697, 329–32; No. 1699, 351–4.

1926   'Les populations du Manyéma (Congo belge)', *Le monde colonial illustré*, Vol. 4, No. 36, 163–4.

1926   'Les neiges de l'Equateur: première ascension du Kilimandjaro par un Français', *L'Illustration*, Vol. 84, No. 4354, 158–9.

1929   'Régions calcaires de l'Indochine', *Annales de géographie*, Vol. 38, No. 213, 266–73.

Louise Rapacka is retired from the Bibliothèque Nationale in Paris, where she was Conservateur of the Service Polonais (Polish Section). Essay translated into English by Mme Rapacka's daughter, Wanda Rapacka of Strasbourg, and edited by Gary Dunbar.

Chronology

1883	Born in Paris, on 13 July
1887	Father, Léon Cuisinier, died in Viarmes, France, on 7 December
1888–90	Lived in Clarens, Switzerland, in home of Elisée Reclus
1890–94	Lived in France with Elisée Reclus, first in Nanterre and then in Paris
1895–97	Lived in Menton
1897	Mother died on 21 March
1897–1901	Lived in Brussels in home of Elisée Reclus
1901–02	Studied mathematics at University of Liège
1902–07	Studied at the University of Geneva in the Faculty of Sciences
1907	First mineral-prospecting mission, in Africa. Made his home base in Geneva until 1931
1910	Married Wanda Tutakiewicz (d. 1942)
1914–18	Interned in Austria
1918	Escaped to Switzerland
1923	Certificate in Geodesy and Astronomy from Army Geographical Service, Paris
1931	Moved to Paris
1940–46	Spent the war years in Indochina, especially in Laos
1952	Died in Paris on 30 August

Index

The index is divided into two parts:

1. A general index, including personal names, organizations, conferences, societies, and geographical concepts, theories and research.

2. A cumulative list of biobibliographies which includes all the geographers listed in volumes 1–16 inclusive.

1. GENERAL INDEX

Academy of Moral and Political Sciences (France) 45, 46
Academy of Sciences, Austria 14
Admiralty handbooks 56
aerial photography 16
Africa 14, 97, 98
agriculture
    Bujak 26
    Meuriot 46, 47
Alchester 54
Alert, HMS 70
algae 70
algebra 97
Alps 13, 97
altitude 33
Alvarez-Brun 83
Amazon river 82
Ammassalik 64
Amundsen, Roald 68
Ancash 81, 82, 83
Andes 81
Angara Basin 72
animals 31, 32
    dissected 81
Annals of Social and Economic History 26
Annual (British School of Athens) 55
Antarctica 67–8, 69
Anthropological Society of London 82
anthropology
    Bujak 23, 24
    Myres 54, 55, 56, 58
    Nansen 71–2

Raimondi del Acqua 81
antiquity
    Myres 54, 55, 56, 57
    Raimondi del Acqua 83
Apurimac river 82
archaeology
    Myres 53, 54, 55, 57
    Raimondi del Acqua 81, 83
Archer, Colin 65
Arctic Ocean 63–6, 67–8, 69–70, 72
Arnberger, Erik 14
Asia Minor, ancient 54
Atlas der Republik Österreich 15
'Atlas of Volcanoes' (Reclus' proposed book) 90
atmospheric sciences 69
Austria 14–15, 18
Austrian Academy of Sciences 14

Badania z dziejów społecznych i gospodarczych 26
Baker, Janet 5–6
Baker, John Norman Leonard 1–11
Baker, Rosalind 5–6
Bakunin, Mikhail 89
Balta, José 85
Ban Ca Tep 98
Barriol, A. 49
Bauer, Walter 75
Beckit, H.O. 2
Beiträge zur Regionalforschung 14
Belgian Society of Astronomy, Meterology and Geophysics 90
Beneventano, Marco 25
Berdoulay, V. 49
Bergen Museum 64

2. CUMULATIVE LIST OF BIOBIBLIOGRAPHIES

AL-BIRUNI (Abu' Rayhan Muhammad) (973–1054)   *13*, 1–9
AL-HASAN ibn Muhammad al-Wazzân az-Zayyâtî, *see* LEO AFRICANUS
ALMAGIÀ, Roberto (1884–1962)   *13*, 11–15
AL-MUQADDASI (*c.* 945–*c.* 988)   *4*, 1–6
ANCEL, Jacques (1882–1943)   *3*, 1–6
ANUCHIN, Dmitry Nikolaevich (1843–1923)   *2*, 1–8
APIANUS, Peter (1495 or 1501–1552)   *6*, 1–6
ARBOS, Philippe (1882–1956)   *3*, 7–12
ARDEN-CLOSE, Charles Frederick (1865–1952)   *9*, 1–13
ARQUÉ, Paul (1887–1970)   *7*, 5–9
ATWOOD, Wallace Walter (1872–1949)   *3*, 13–18
AUROUSSEAU, Marcel (1891–1983)   *12*, 1–8

BAKER, John Norman Leonard (1893–1971)   *16*, 1–11.
BANSE, Ewald (1883–1953)   *8*, 1–5
BARANSKIY, Nikolay Nikolayevich (1881–1963)   *10*, 1–16
BATES, Henry Walter (1852–1892)   *11*, 1–5
BAULIG, Henri (1877–1962)   *4*, 7–17
BERG, Lev Semenovich (1876–1950)   *5*, 1–7
BERNARD, Augustin (1865–1947)   *3*, 19–27
BINGHAM, Millicent Todd (1880–1968)   *11*, 7–12
BLACHE, Jules (1893–1970)   *1*, 1–8
BLODGET, Lorin (1823–1901)   *5*, 9–12
BOBEK, Hans (1903–1990)   *16*, 12–22
BOSE, Nirmal Kumas (1901–1972)   *2*, 9–11
BOWEN, Emrys George (1900–1983)   *10*, 17–23
BOWMAN, Isaiah (1878–1950)   *1*, 9–18
BRATESCU, Constantin (1882–1945)   *4*, 19–24
BRAWER, Abraham Jacob (1884–1975)   *12*, 9–19
BRIGHAM, Albert Perry (1855–1929)   *2*, 13–19
BROOKS, Alfred Hulse (1871–1924)   *1*, 19–23
BROWN, Ralph Hall (1898–1948)   *9*, 15–20
BROWN, Robert Neal Rudmose (1879–1957)   *8*, 7–16
BUACHE, Philippe (1700–1773)   *9*, 21–7
BUJAK, Franciszek (1875–1953)   *16*, 23–30
BUSCHING, Anton Friedrich (1724–1793)   *6*, 7–15

CAMENA d'ALMEIDA, Pierre (1865–1943)   *7*, 1–4